TWENTY FIRST CENTURY
SCI⊙nce

GCSE Additional Science

Project Directors

Jenifer Burden   Andrew Hunt

John Holman   Robin Millar

Project Officers

Peter Campbell   Angela Hall

John Lazonby   Peter Nicolson

Course Editors

Jenifer Burden   Andrew Hunt   Robin Millar

Authors

Biology	Chemistry	Physics
Jenifer Burden	Anna Grayson	Simon Carson
Bill Indge	John Holman	Robin Millar
Jean Martin	Andrew Hunt	Stephen Pople
Nick Owens	John Lazonby	Carol Tear
Lynn Winspear	Allan Mann	

OCR
RECOGNISING ACHIEVEMENT

Nuffield
Curriculum Centre

THE UNIVERSITY of York

OXFORD

Contents

Introduction

Welcome to *Twenty First Century Science*

GCSE Additional Science

This course looks more deeply at science explanations. As you go through the course, you will develop a greater understanding of the world around you. By the end of it, you will be well prepared to study any science course after GCSE.

But this is a course for anyone who is interested in science, whether or not you choose a science career. Studying science teaches you how to interpret data, think logically, and communicate clearly and carefully. These are all valuable skills which are needed in many different jobs that are not science based.

How to use this book

The book is divided into nine modules. Each module is about a different topic: three modules look at biology (B4–B6), three at chemistry (C4–C6), and three at physics (P4–P6).

If you want to find a particular area of science, use the **Contents** and **Index** pages. You can also use the **Glossary**. This explains all the key words used in the book.

Each module has two introduction pages, which tell you the main ideas you will study. They look like this:

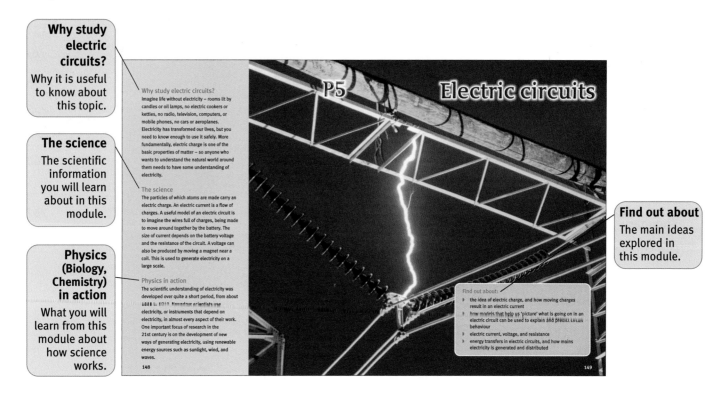

Why study electric circuits?
Why it is useful to know about this topic.

The science
The scientific information you will learn about in this module.

Physics (Biology, Chemistry) in action
What you will learn from this module about how science works.

Find out about
The main ideas explored in this module.

Why study electric circuits?
Imagine life without electricity – rooms lit by candles or oil lamps, no electric cookers or kettles, no radio, television, computers, or mobile phones, no cars or aeroplanes. Electricity has transformed our lives, but you need to know enough to use it safely. More fundamentally, electric charge is one of the basic properties of matter – so anyone who wants to understand the natural world around them needs to have some understanding of electricity.

The science
The particles of which atoms are made carry an electric charge. An electric current is a flow of charges. A useful model of an electric circuit is to imagine the wires full of charges, being made to move around together by the battery. The size of current depends on the battery voltage and the resistance of the circuit. A voltage can also be produced by moving a magnet near a coil. This is used to generate electricity on a large scale.

Physics in action
The scientific understanding of electricity was developed over quite a short period, from about 1800 to 1910. Nowadays scientists use electricity, or instruments that depend on electricity, in almost every aspect of their work. One important focus of research in the 21st century is on the development of new ways of generating electricity, using renewable energy sources such as sunlight, wind, and waves.

P5 Electric circuits

Find out about:
▶ the idea of electric charge, and how moving charges result in an electric current
▶ how models that help us 'picture' what is going on in an electric circuit can be used to explain and predict circuit behaviour
▶ electric current, voltage, and resistance
▶ energy transfers in electric circuits, and how mains electricity is generated and distributed

148

149

Each module is split into sections. Pages in a section look like this:

Heading
Each section looks at a different part of the module.

Find out about
The key points explored in the section.

Higher Tier
The 'H' flag next to something on the page means that it refers to Higher Tier material in the specification.

Example
Examples boxes show you how to do calculations.

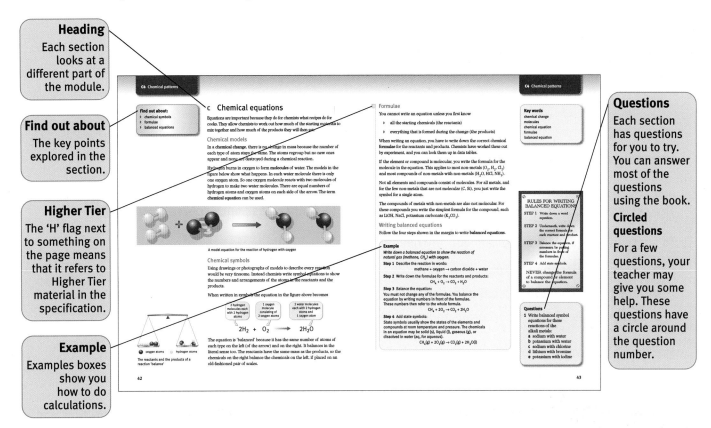

Questions
Each section has questions for you to try. You can answer most of the questions using the book.

Circled questions
For a few questions, your teacher may give you some help. These questions have a circle around the question number.

Each module ends with a summary and some questions. They look like this:

Summary
A checklist of key points explained in this module.

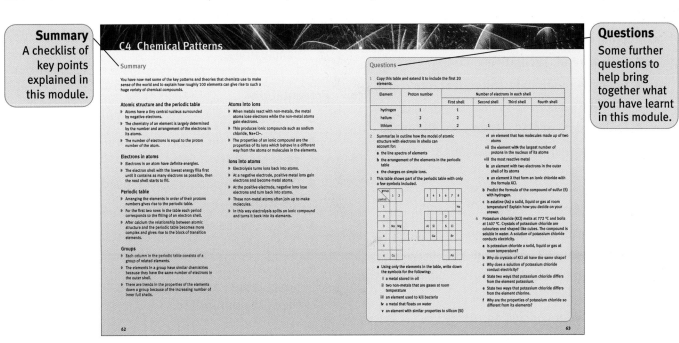

Questions
Some further questions to help bring together what you have learnt in this module.

Internal assessment

In GCSE Additional Science your internal assessment counts for 33.3% of your total grade. Marks are given for a practical investigation.

Your school or college may give you the marking schemes for this. This will help you understand how to get the most credit for your work.

Internal assessment (33.3% of total marks)

Investigation (33.3%)

Investigations are carried out by scientists to try and find the answers to scientific questions. The skills you learn from this work will help prepare you to study any science course after GCSE.

To succeed with any investigation you will need to:

▶ choose a question to explore

▶ select equipment and use it appropriately and safely

▶ design ways of making accurate and reliable observations

Marks will be awarded under five different headings.

Strategy

▶ Choose the task for your investigation.

▶ Decide how much data you need to collect.

▶ Choose equipment and techniques to give you precise and reliable data.

Collecting data

▶ Take careful, accurate measurements safely.

▶ Collect enough data and repeat it to check reliability.

▶ Collect data across a wide enough range.

▶ Control other things which might affect your results.

Interpreting data

▶ Use charts, tables, diagrams, or graphs to make clear any patterns in your results.

▶ Say what conclusions you can make from your data.

▶ Explain your conclusions using your science knowledge and understanding.

Evaluation

▶ Look back at your experiment and say how you could improve the method.

▶ Explain how reliable you think your evidence is.

▶ Suggest some improvements or extra data you could collect to be more confident in your conclusions.

Presentation

▶ Write a full report of your investigation.

▶ Make sure your report is laid out clearly in a sensible order.

▶ Describe the apparatus you used and what you did with it.

▶ Remember to show all units correctly (e.g. cm^3).

▶ Take care with your spelling, grammar, and punctuation, and use scientific terms where they are appropriate.

Secondary data

Your investigation report will be based on the data you collect from your own experiments. You may also use information from other people's work. This is called secondary data. You could **get** secondary data from other students in your class. You may also be able to find data from other sources. For example:

- ▶ Internet
- ▶ school library
- ▶ local public library
- ▶ your science textbook
- ▶ interview a scientist
- ▶ write a letter to an organization
- ▶ telephone to find out who to write to

To get useful information from other people, make sure you have detailed questions beforehand.

Speak (or write) politely and explain who you are and what you are doing.

Make sure you ask for just the information you really need.

When will you do this work?

Your investigation must be based on practical work you do in class. Your school or college will decide when you do your internal assessment. If you do more than one investigation, they will choose the best one for your marks.

Tip

The best advice is 'plan ahead'. Give your work the time it needs and work steadily and evenly over the time you are given. Your deadlines will come all too quickly, especially as you will have coursework to do in other subjects.

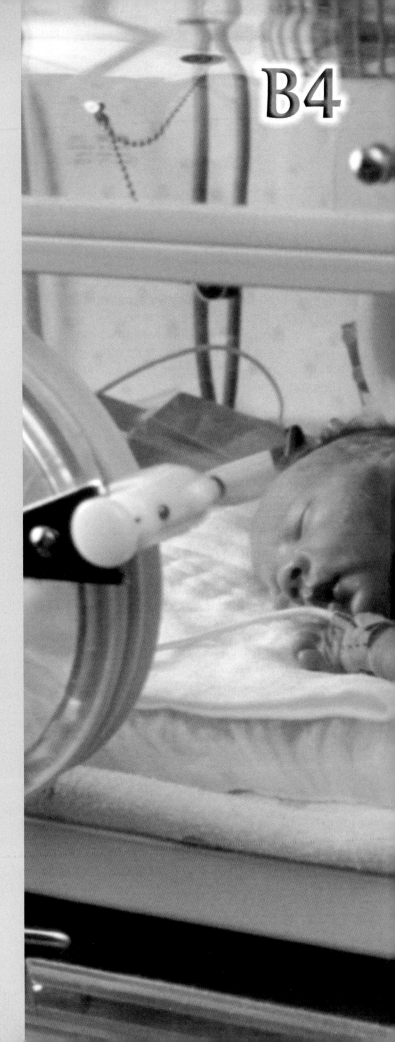

Why study homeostasis?

Every moment of your life your body is reacting to changes. Some of these changes happen outside your body. Others happen inside your body, for example your body's water level dropping. Your body responds to these changes to make sure that conditions inside your body stay steady. This is vital for survival.

The science

Keeping a steady state inside your body is called homeostasis. Automatic systems in the body control water balance and body temperature. Water molecules are constantly moving in and out of cells. The amount of heat you lose to your environment depends on several factors. For example, is it a warm or a cold day? Enzymes speed up chemical reactions in cells. They need a particular temperature to work at their fastest rate.

Biology in action

Some diseases damage the body's ability to keep conditions inside steady. Extreme environments and sports can put too much strain on the body, and homeostasis starts to fail. Understanding homeostasis is crucial to help many people, from a baby in an incubator, to a person with kidney disease.

Homeostasis

Find out about:

- why homeostasis is important for your cells
- how temperature affects your enzymes
- how your body temperature is kept constant
- how different chemicals move in and out of your cells
- how your kidneys control your body's water level

Find out about:
▶ homeostasis
▶ why it is important

A Changing to stay the same . . .

Inside your cells thousands of chemical reactions are happening every second. These reactions are keeping you alive. But for your cells to work properly they need certain conditions. Keeping conditions inside your body the same is called **homeostasis**.

Homeostasis is not easy – lots of things have to happen for your body to 'stay the same'. Look at just a few of the changes happening every second:

Running has made this athlete hot. His body is sweating more to cool back down. This is an example of homeostasis.

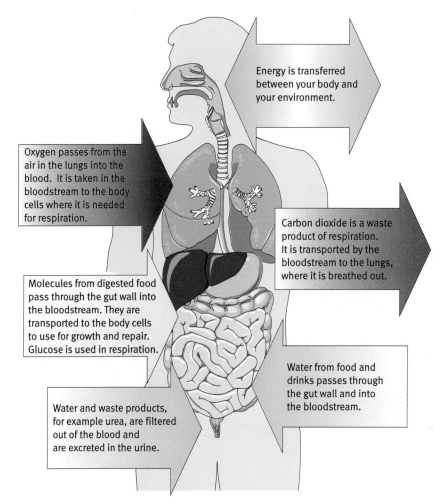

Energy is transferred between your body and your environment.

Oxygen passes from the air in the lungs into the blood. It is taken in the bloodstream to the body cells where it is needed for respiration.

Carbon dioxide is a waste product of respiration. It is transported by the bloodstream to the lungs, where it is breathed out.

Molecules from digested food pass through the gut wall into the bloodstream. They are transported to the body cells to use for growth and repair. Glucose is used in respiration.

Water from food and drinks passes through the gut wall and into the bloodstream.

Water and waste products, for example urea, are filtered out of the blood and are excreted in the urine.

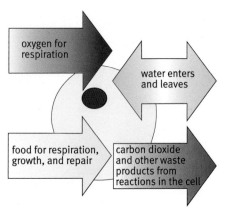

oxygen for respiration

water enters and leaves

food for respiration, growth, and repair

carbon dioxide and other waste products from reactions in the cell

Cells must take in materials for respiration, growth, and repair. They must also get rid of waste.

Some of the inputs and outputs that are going on all the time in your body. Materials enter and leave. Energy is also transferred in or out along a temperature gradient. The direction of transfer depends on the temperature of the environment.

So your body must work hard to:

▶ keep a constant body temperature

▶ keep the correct levels of water and salt

▶ control the amounts of nutrients, for example glucose

▶ take in enough oxygen for respiration

▶ get rid of toxic waste products, for example carbon dioxide and urea

Control systems

Premature babies cannot control their temperature, so they are put in incubators. The incubator is an artificial control system.

The control systems keeping a steady state in your body work in a similar way to artificial control systems.

All control systems have:

▶ a **receptor**, which detects the stimuli (the change)

▶ a **processing centre**, which receives the information

▶ an **effector**, which produces an automatic response

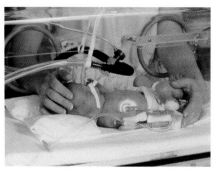

An artificial control system is keeping this baby's temperature steady.

How does the incubator work?

An incubator has a temperature sensor, a thermostat with a switch, and a heater.

The temperature inside the incubator is detected by the sensor. If the temperature falls, the thermostat switches on the heater. When the temperature gets back up to its normal setting of 32 °C, the thermostat switches off the heater.

What about your body?

Some of the temperature control in your body is automatic too. For example, you do not consciously decide to sweat when you are hot. But you can also consciously control your temperature by doing things like putting a coat on or taking it off, or moving between a shady and a sunny place.

There are temperature receptors in your brain and skin. If you are cold, you can warm yourself by putting on a jacket.

Questions

1 Write down a definition for homeostasis.

2 In an incubator, name:

  a a receptor

  b a processing centre

  c an effector

3 You can cool yourself down by taking a coat off. In this control system, name:

  a the receptors

  b the processing centre

  c the effectors

4 Explain why the temperature in an incubator is set below 37 °C even though the baby's body must stay at this temperature.
(*Hint:* Respiration in the baby's cells will be releasing energy.)

Key words

homeostasis
receptor
processing centre
effector

Find out about:

▶ negative feedback
▶ how some effectors work in opposite pairs

B Feedback in control systems

If the temperature in an incubator falls too low, the heater is switched on. The temperature goes up. When the temperature is high enough, the heater is switched off. This type of control is called **negative feedback**:

▶ Any change in the system results in an action that reverses the change.

The diagram below explains how this works.

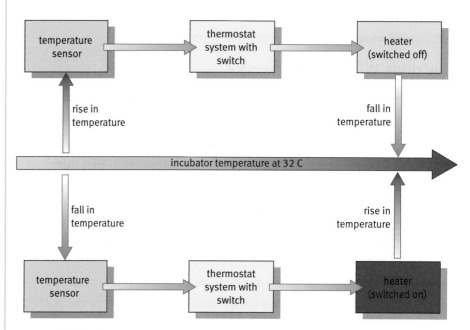

Negative feedback in an incubator to control temperature

Negative feedback systems are all around you. For example, if the temperature inside your fridge goes up, the motor switches on to cool it down. When it is cool enough, the motor switches off.

Your body uses negative feedback systems too. However, your body systems are more complicated than those in an incubator or a fridge. You have effectors to cool you down *and* to warm you up.

Control systems with more than one effector

Imagine you are driving a car in a 30 mph zone. You need to make sure that you do not speed, but you should not drive too slowly either. In this system, your brain is the processing centre. There are two effectors: an accelerator and a brake.

Now imagine you could have only one of those effectors, and you are going down a hill.

  ◗ If you only have a brake, you will be fine – until you get to the bottom.

  ◗ If you only have an accelerator, you have problems! Taking your foot off the accelerator is not going to be enough to control your speed.

You need both of these effectors to have proper control of the car. The brake and accelerator work together to maintain a constant speed. When the car speeds up, you brake. If the speed falls too low, you use the accelerator to speed it up again.

Because they have opposite effects, the brake and accelerator are called **antagonistic effectors**. You can use them to adjust the system as soon as there is a change – either too fast or too slow. This means that the response is much more sensitive. Your body uses antagonistic effectors too.

You need both an accelerator and a brake to control a car's speed.

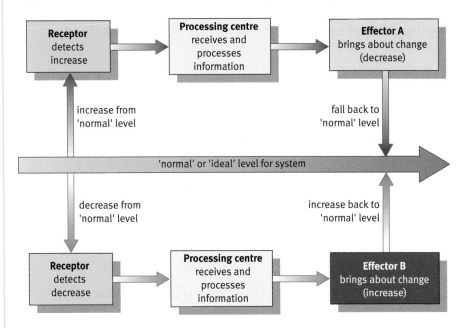

A control system with two effectors that have an opposite, or antagonistic, effect is more sensitive than a system with only one effector.

Key words

negative feedback
antagonistic effectors

Questions

1 Write down a definition for negative feedback.

2 If the diagram above represents a car, what is:

  a effector A?   **b** effector B?

3 Why are the brake and accelerator of a car called antagonistic effectors?

4 Why is it an advantage to have antagonistic effectors in a car?

5 Central heating systems use negative feedback to control the temperature of a house. Explain simply how this works.

Find out about:
- why you cannot live without enzymes
- how enzymes work
- how temperature and pH effect enzymes

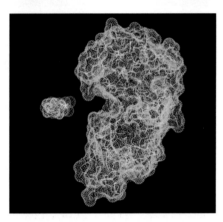

A computer graphic of an enzyme, its active site, and the product of a reaction.

c Enzymes

The chemical reactions that take place in cells rely on **enzymes**. Enzymes work best in certain conditions, for example at a certain temperature. This is an important reason why you need a steady state inside your cells.

What are enzymes?

Enzymes are the **catalysts** that speed up chemical reactions in living organisms. They are proteins – large molecules made up of long chains of amino acids. The amino acid chains are different in each protein, so they fold up into different shapes. An enzyme's shape is very important to how it works.

How do enzymes work?

Some enzymes break down large molecules into smaller ones. Others join small molecules together.

In all cases, the molecules must fit exactly into a part of the enzyme called the **active site**. It is a bit like fitting the right key in a lock. So scientists call the explanation of how an enzyme works the **lock-and-key model**.

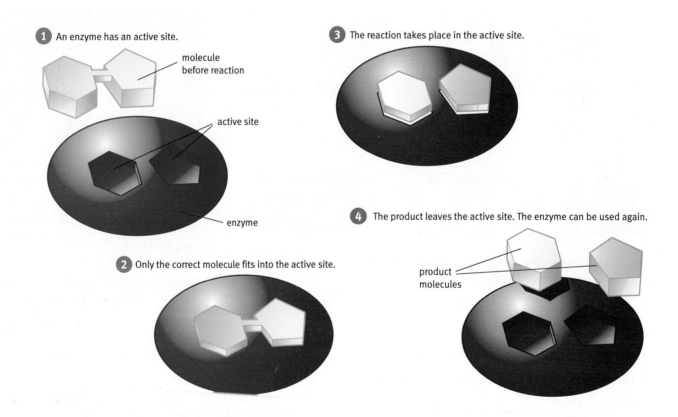

1 An enzyme has an active site.

molecule before reaction

active site

enzyme

2 Only the correct molecule fits into the active site.

3 The reaction takes place in the active site.

4 The product leaves the active site. The enzyme can be used again.

product molecules

The lock-and-key model of enzyme function. (*Note:* This diagram is schematic. This means that if you were able to see the molecules and enzymes, they would not look like they do here. Enzymes consist of many molecules – see the photograph above left for an idea of scale. However, this is a useful way of picturing, or modelling, how an enzyme works.)

Why do we need enzymes?

At 37 °C, chemical reactions in your body would happen too slowly to keep you alive.

One way of speeding up a reaction is to increase the temperature. As the temperature rises, molecules

- ▶ have more energy
- ▶ move around faster and collide more often
- ▶ react more easily when they do collide

A higher body temperature could speed up the chemical reactions in your body. But higher temperatures damage human cells. Also, to keep your body warm, you have to release energy from respiration. For a higher body temperature, you would need a lot more food to fuel respiration.

So, we rely on enzymes to give us the rates of reaction that we need. They can increase rates of reaction by up to 10 000 000 000 times. It is not possible to live without enzymes.

If you had to maintain a higher body temperature, you would have to spend a lot more time eating to provide the calories to heat your body. Already 80% of the energy from your food is used for keeping warm.

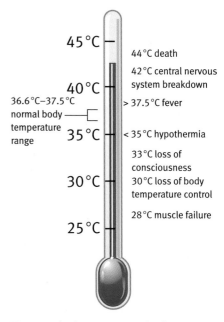

Your core body temperature is about 37 °C, but a small variation either side of this is normal. A core temperature over 42 °C or under 28 °C usually results in death.

Shrews have a large surface area for their volume. They lose heat to the environment over their whole body surface. To release enough energy to maintain their body temperature, they have to eat 75% of their body mass in food each day. If not, they die within 2–3 hours.

Questions

1 Write down:

   a what enzymes are made of   **b** what enzymes do

2 Explain how an enzyme works. Use the key words on this page in your answer.

3 The enzyme amylase breaks down starch to sugar (maltose); catalase breaks down hydrogen peroxide to water and oxygen. Explain why catalase does not break down starch.

④ **a** Calculate from your own body mass how much food you would need to consume each day if you had to eat the same proportion of your body mass as a shrew.

   b How many 500 g loaves of bread does this mass of food represent?

Key words

enzymes
catalysts
lock-and-key model
active site

Why do higher temperatures speed up reactions?

For any reaction to happen, the molecules must bump into each other. At low temperatures, molecules move slowly, so they are less likely to collide. If the temperature increases, the molecules have more energy. They move around faster. The molecules collide more often and with more energy. So the rate of reaction increases.

How does temperature affect enzymes?

At low temperatures, enzyme reactions get faster if the temperature is increased. The enzymes and other molecules collide more often, so they react more frequently.

But there is a difference with enzyme reactions. Above a certain temperature the reaction stops. This is because enzymes are proteins. Higher temperatures change an enzyme's shape so that it no longer works. The diagram below explains what happens using the lock-and-key model.

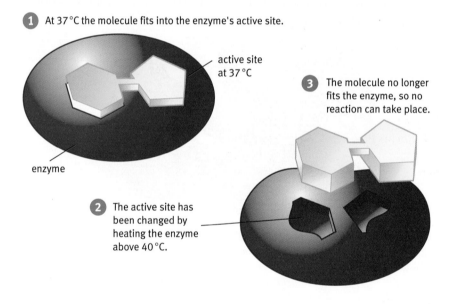

1 At 37 °C the molecule fits into the enzyme's active site.

active site at 37 °C

enzyme

3 The molecule no longer fits the enzyme, so no reaction can take place.

2 The active site has been changed by heating the enzyme above 40 °C.

How an enzyme reaction can be stopped by a rise in temperature.

'Ice fish' such as Antarctic cod are active at 2 °C because their enzymes work best at low temperatures. Most organisms would be dead or very sluggish at 2 °C.

Bacteria living in hot springs have enzymes that withstand high temperatures.

Enzymes are denatured at high temperatures

High temperatures change the shape of an enzyme. They do not destroy it completely. But even when an enzyme cools down, it does not go back to its original shape. Like cooked egg white, the protein cannot be changed back. The enzyme is said to be **denatured**.

Why 37 °C?

The temperature at which an enzyme works best is called its **optimum temperature**. It is a temperature too low to denature the enzyme. But it is high enough for collisions between the enzyme and other molecules to be frequent and energetic. Enzymes in humans work best at around 37 °C. Some organisms have cells and enzymes adapted to different temperatures.

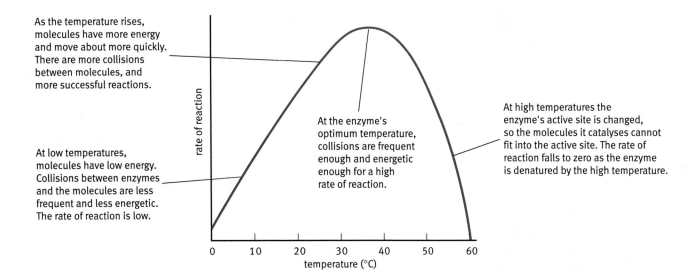

As the temperature rises, molecules have more energy and move about more quickly. There are more collisions between molecules, and more successful reactions.

At low temperatures, molecules have low energy. Collisions between enzymes and the molecules are less frequent and less energetic. The rate of reaction is low.

At the enzyme's optimum temperature, collisions are frequent enough and energetic enough for a high rate of reaction.

At high temperatures the enzyme's active site is changed, so the molecules it catalyses cannot fit into the active site. The rate of reaction falls to zero as the enzyme is denatured by the high temperature.

The optimum temperature of an enzyme

pH also affects enzymes

The shape of an enzyme's active site is also affected by pH. Every enzyme has an optimum pH at which it works best.

> **Key words**
> denatured
> optimum temperature

Enzyme	What it does	Optimum pH
salivary amylase	breaks down starch to sugar (maltose)	4.8
pepsin	breaks down proteins into short chains of amino acids	2.0
catalase	breaks down hydrogen peroxide into water and oxygen	7.6

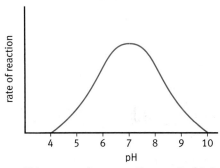

This enzyme has an optimum pH of 7.0.

Questions

5 Explain why increasing the temperature causes enzyme reactions to:

  a get faster at low temperatures

  b stop at higher temperatures

6 What is meant by an enzyme's optimum temperature?

7 Amylase is an enzyme that changes starch to maltose. Some students compared the rate of reaction at 17 °C, 37 °C, and 57 °C.

  a Say what would happen if the students:

  i raised the temperature of the tube at 17 °C to 37 °C?

  ii lowered the temperature of the tube at 57 °C to 37 °C?

  b Explain both of these changes.

8 Enzymes in food and enzymes from microorganisms such as bacteria and fungi make food decay. Why does food stay fresh in a refrigerator for a few days but for months in a freezer?

9 Amylase is in your saliva and your small intestines. Suggest why amylase cannot work in your stomach.

Find out about:
▸ how you warm up and cool down

D Getting hot, getting cold!

Imagine you are on a tropical island in summer. You are lying on the beach. The sand is baking hot. You slowly drift off to sleep . . . but inside, your body is working hard to keep a constant body temperature.

Gaining heat

If your environment is hotter than you are, energy will be transferred to your body. Your body also gains heat from **respiration**. During respiration, glucose is broken down to release energy for cells. This energy is used by muscles for movement. And some of the energy warms up the body.

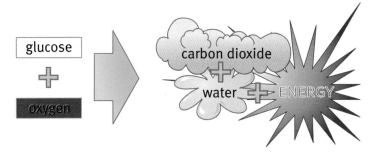

Respiration releases energy from the breakdown of glucose.

Losing heat

If your environment is cooler than you are, energy will be transferred away from your body. The bigger the temperature difference, the greater the rate of cooling. The swimmers on the left are losing energy to the sea and air.

Getting the balance right

For your temperature to remain constant, energy gain must be balanced by energy loss.

In other words, if heat gained = heat lost, your body temperature stays the same.

Different parts, different temperatures

Not all of your body is at the same temperature. Your **extremities** (hands and feet) are cooler than your **core** (deeper parts). They have a larger surface area compared with their size. So they lose energy to the environment faster than the main parts of your body.

The temperature of your hands, feet, and skin falls, but your core temperature hardly changes. Your core body temperature is the temperature inside your trunk and your brain. It should be between 36 °C and 37.5 °C for your body to work properly. 'Normal' body temperature will vary from one person o the next.

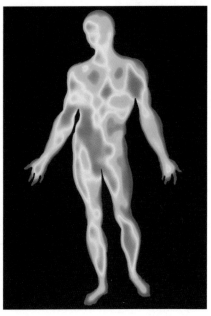

Luckily, these swimmers' bodies respond quickly to their fall in temperature to bring it back to normal.

On this thermal image, the hottest parts of the body are red and the coolest, blue.

The amount of energy released in respiration and other reactions is greatest in your muscles and liver. The circulation of your blood transfers this energy to other parts. When you are cold, the circulation to your extremities is reduced. This means that your core stays warm.

Investigating temperature control

A physiologist called Sir Charles Blagden was Secretary of the Royal Society towards the end of the eighteenth century. Like many physiologists, he experimented on himself. He went into a very hot room to see how his body would react. The account below is adapted from Harry Houdini's book, *The Miracle Mongers: An Exposé*.

Another account describes Blagden taking a dog into the room with him, along with steak and eggs. The dog was unharmed but had to stay in its basket so it did not burn its feet on the hot floor.

> **Key words**
> respiration
> extremities
> core

Blagden's experiment

Sir Charles Blagden went into a room where the temperature was 1 degree or 2 degrees above 127 °C, and remained eight minutes in this situation, frequently walking about to all the different parts of the room, but standing still most of the time in the coolest spot, where the temperature was above 116 °C. The air, though very hot, gave no pain, and Sir Charles and all the other gentlemen were of the opinion that they could support a much greater heat.

During seven minutes Sir Charles' breathing continued perfectly good, but after that time he felt an oppression in his lungs, with a sense of anxiety, which induced him to leave the room. His pulse was then 144, double its ordinary quickness. In order to prove that the thermometer was not faulty, they placed some eggs and a beef-steak upon a tin frame near the thermometer, but more distant from the furnace than from the wall of the room. In twenty minutes the eggs were roasted quite hard, and in forty-seven minutes the steak was not only cooked, but almost dry.

Blagden's dog takes part in the experiment.

Questions

1 What two things must be balanced to keep your body temperature balanced?

2 What part of your body is warmest?

3 What temperature does this warmest part need to be kept at?

4 Why is your blood important in keeping extremities warm?

5 What effects did the very high temperatures have on Sir Charles Blagden?

6 What happened to the proteins in the steak and eggs as they cooked?

7 Why did the same thing not happen to Sir Charles' proteins?

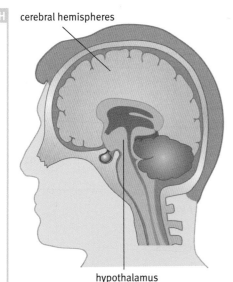

cerebral hemispheres

hypothalamus

The **hypothalamus** is the processing centre in the brain for sleep, water balance, body temperature, appetite, and other functions. The cerebral hemispheres are where you make conscious decisions to warm or cool yourself.

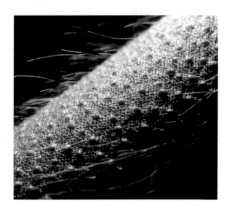

Goose pimples are the result of contraction of the muscles in your skin that make your hair stand on end. The insulating layer of air trapped by hair is effective in furry mammals.

Sensing and control

Changes in your body temperature can have serious effects on health. So temperature receptors in your skin can detect a change in air temperature of as little as 0.5 °C. Your brain is particularly sensitive to temperature changes. So there are receptors here that detect blood temperature. Your brain also contains the processing centre for temperature control. It receives information from the temperature receptors. When the temperature in your brain is above or below 37 °C, it automatically triggers effectors to bring your body temperature back to normal. These effectors include muscles and sweat glands.

Warming up

Shivering is one way your body keeps warm. When you shiver, muscle cells contract quickly. They must respire faster to release the energy for this movement. So there is also more energy for keeping warm.

Shivering is an automatic response. You may also take a conscious decision to do something that will warm you up, for example drinking a warm drink, putting on more clothes, or going inside.

Cooling down

When you are too hot, nerve impulses from the brain stimulate your **sweat glands**. They make sweat, which passes out of small pores onto the skin surface. When sweat **evaporates** from your skin, it cools you down. Even when it is cool and you are not very active, you can lose nearly a litre of water a day in sweat. When you are hot and active, you can lose up to three litres of water an hour. That amount of water is hard to replace, so you could become dehydrated.

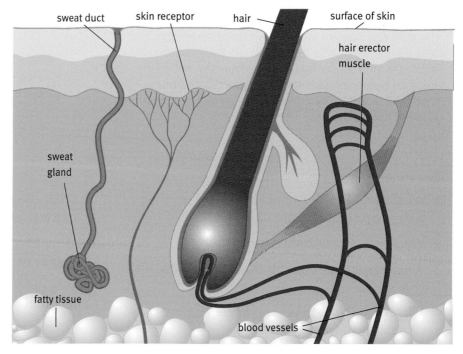

The structure of skin

How does sweat cool you down?

Water molecules in sweat gain energy from your skin. Soon they move quickly enough to evaporate. This cools you down.

Sweating only works to cool you down if the sweat can evaporate quickly. In a hot, humid climate sweat drips off you and you feel uncomfortably hot. Air currents increase the rate of evaporation and so increase the cooling effect.

Is your body temperature the same all day?

All these responses keep your body temperature within a narrow range. Average core body temperature is about 36. 9 °C. But this varies from person to person and it varies throughout the day. For example, when you are sleeping, you move around less and respire more slowly. So your body temperature drops to its lowest point at night.

In hot, humid conditions, sweat cannot evaporate easily. Your clothes may become soaked in sweat.

The fans in this Sikh temple provide a welcome breeze to help to keep the people cool.

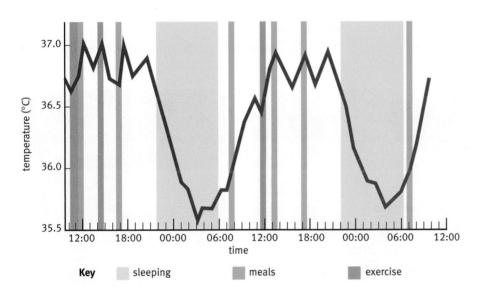

The daily cycle of variation in body temperature. You can see that eating, sleeping and activity affect body temperature. But these fluctuations happen when you are at rest too – they are controlled by our 'biological clocks'.

Questions

8 Where in your body would you find:
a temperature receptors?
b the temperature processing centre?

9 Name two effectors for controlling body temperature.

10 Explain how shivering warms you up.

11 Explain how sweating cools you down.

Key words
hypothalamus
shivering
sweat glands
evaporates

Too hot or too cold?

The table below lists ways ways of controlling body temperature.

Too cold?	Too hot?
Shivering	**Sweating (or getting wet)**
Energy warms the body tissues when muscles contract. This also happens when you exercise.	Energy is lost from the skin molecules when the water in sweat evaporates.
Warm food or drinks	**Cold food or drinks**
Energy is transferred from the warm food or drink to your body.	You are cooled as energy from your body heats the cold food or drink.
Clothes and hair-raising	**Protective clothing**
Clothes and hair trap an insulating layer of air. This slows the loss of energy from your body to the environment.	Hats, sun umbrellas, and protective clothing can reduce the heating effect of direct sunlight. An insulating layer can keep you cool in hot weather.
Heater	**Fan**
Go into an area that is warmer than your body, for example near a heater or in the sun. Energy is transferred to your body from the environment.	Moving air near the skin can increase the rate of evaporation of sweat. This increases the cooling effect.
Vasoconstriction H	**Vasodilation** H
Loss of energy from your body's surface is reduced.	Energy loss from your body's surface is increased.

A

B

C

D

E

F

Questions

12 For each picture on this page, say why the action cools the body.

Vasoconstriction and vasodilation

The boy on the right is flushed as a result of **vasodilation**. 'Vaso' is from the Latin for vessel, so vasodilation means widening of the blood vessels. More blood flows into the capillaries in the skin, so there is more energy transfer to the environment. The opposite is **vasoconstriction**. Less blood reaches the capillaries in the skin, so energy loss is reduced.

This boy has a fever. He is sweating and looking flushed.

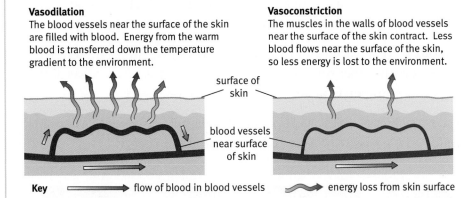

Vasodilation
The blood vessels near the surface of the skin are filled with blood. Energy from the warm blood is transferred down the temperature gradient to the environment.

Vasoconstriction
The muscles in the walls of blood vessels near the surface of the skin contract. Less blood flows near the surface of the skin, so less energy is lost to the environment.

surface of skin

blood vessels near surface of skin

Key ⟶ flow of blood in blood vessels   ⟿ energy loss from skin surface

Key words
vasodilation
vasoconstriction

Vasodilation and vasoconstriction are a good example of the effects of control by antagonistic effectors.

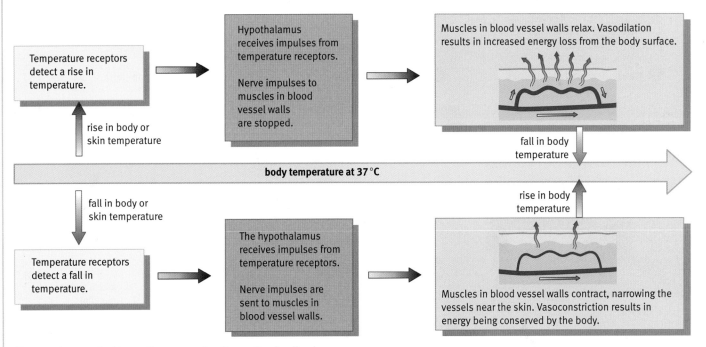

Temperature receptors detect a rise in temperature.

rise in body or skin temperature

Hypothalamus receives impulses from temperature receptors.

Nerve impulses to muscles in blood vessel walls are stopped.

Muscles in blood vessel walls relax. Vasodilation results in increased energy loss from the body surface.

fall in body temperature

body temperature at 37 °C

fall in body or skin temperature

rise in body temperature

Temperature receptors detect a fall in temperature.

The hypothalamus receives impulses from temperature receptors.

Nerve impulses are sent to muscles in blood vessel walls.

Muscles in blood vessel walls contract, narrowing the vessels near the skin. Vasoconstriction results in energy being conserved by the body.

Temperature control is another example of negative feedback.

Questions

13 What are the effectors that cause vasodilation and vasoconstriction?

14 Explain why vasodilation and vasoconstriction are said to be controlled by antagonistic effectors.

⑮ Why does vasodilation not cool you when the temperature of the environment is higher than your body temperature?

Find out about:

▶ how chemicals move in and out of cells

▶ why cells need a steady water balance

E Diffusion and osmosis

Your cells need a constant supply of raw materials for chemical reactions. Waste products must also be removed from cells. So molecules move in and out of cells all the time.

Diffusion

Molecules in gases and liquids move about randomly. They collide with each other and change direction. This makes them spread out.

Overall, more molecules move away from where they are concentrated than move the other way. We say the molecules diffuse from areas of their high concentration to areas of low concentration.

For example, molecules diffuse out of a tea-bag when you make a cup of tea. Diffusion is a **passive** process. It does not need any energy.

① Water just poured onto tea-bag

② About 30 seconds later

③ About 2 minutes later

Molecules diffusing out of a tea-bag.

This swimmer's cells need oxygen and glucose for respiration. They must get rid of carbon dioxide. These molecules move in and out of cells by diffusion.

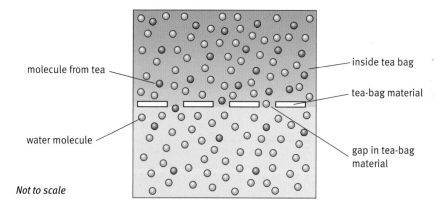

Not to scale

molecule from tea
water molecule
inside tea bag
tea-bag material
gap in tea-bag material

Dissolved molecules from the tea are more concentrated in the tea-bag than in the surrounding water. As the tea brews, more of the dissolved molecules move out of the tea-bag than in. More water molecules will move into the tea bag than out. Eventually, the tea molecules will be evenly spread. (*Note:* In this diagram the circles represent molecules, not individual atoms. The tea-bag material is also made of molecules.)

Cell membranes are partially permeable

The gaps in a tea bag allow water and dissolved molecules to diffuse in and out freely. The bag is permeable. Cell membranes let some molecules through, but not others. We call them **partially permeable membranes**.

The diagrams below show partially permeable membranes. These membranes let water through but not glucose.

Key

partially permeable membrane allows some molecules through and acts as a barrier to others

⚪ glucose molecule

○ water molecule

water molecules associated with glucose molecule

(*Note*: In these diagrams, the circles represent molecules, not individual atoms. Cell membranes are also made of molecules.)

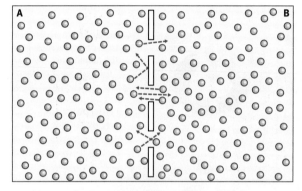

Not to scale

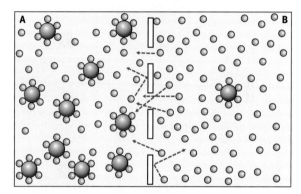

a This membrane is separating water molecules. Water molecules move at random. So as many pass from left (A) to right (B) as pass from right (B) to left (A).

b This membrane is separating two glucose solutions. There are more water molecules and fewer glucose molecules on the right (B). In other words, water is in higher concentration on the right. So there is overall movement of water from right (B) to left (A).

Osmosis

A solution that has a high concentration of water molecules is a dilute solution. For example, if you make a dilute drink of squash, it has lots of water in it. A more concentrated drink of squash would have less water.

More water molecules move away from an area of higher concentration of water molecules. Think of it as diffusion of water. This overall flow of water from a dilute to a more concentrated solution across a partially permeable membrane is called **osmosis**.

Key words
diffusion
partially permeable membrane
osmosis

Questions

1 Write down a definition for diffusion.

2 Name three chemicals that move in and out of cells by diffusion.

3 Explain how diffusion lets you detect a fish and chip shop when you are still around the corner.

4 Explain what is meant by a partially permeable membrane.

5 Write down a definition of osmosis.

6 Draw a diagram like the ones above showing an overall movement of water molecules from solution A to solution B.

Another way of getting molecules into cells

Remember that cell membrancs are partially permeable. Some molecules that the cells need cannot diffuse through membranes into cells. Others diffuse through very slowly.

Sometimes a cell needs to take in molecules that are in higher concentration inside the cell than outside. The molecules cannot move by diffusion.

So cells have another way of moving molecules in or out. It is called **active transport**. Cells use the energy from respiration to transport molecules across the membrane. Glucose is one chemical that is moved into cells by active transport.

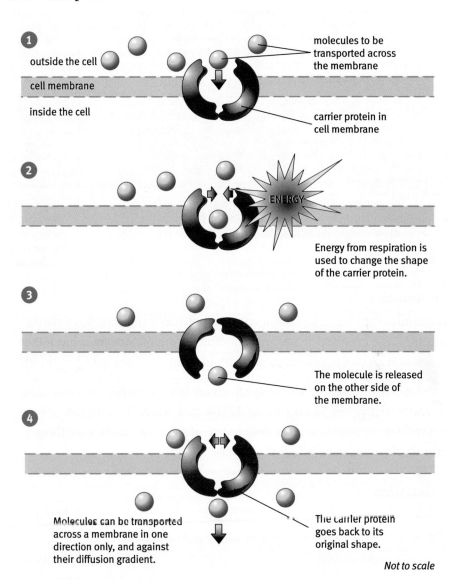

Movement of molecules across a cell membrane by active transport (schematic).

Questions

7 Write down two main differences between diffusion and active transport.

8 Name one chemical which is moved into cells by active transport.

9 Explain why cells sometimes need to use active transport.

Keeping the balance

In most animals that live in the sea, the concentration of dissolved chemicals in their cells and body fluids is about the same as that of seawater. So they do not have a problem with water balance. Water molecules enter and leave their bodies at the same rate.

Freshwater animals do have a problem. The concentration of dissolved chemicals in their bodies is much higher than in freshwater. Water constantly enters their bodies by osmosis. Freshwater animals, such as *Paramecium* (see the diagram on the right), have to use energy to pump water out of their body.

Animals that live on land have the opposite problem. They lose water to the environment all the time, for example in sweating, panting, and urination. So land animals must replace water all the time. If they do not, their body fluids become too concentrated and their cells do not work properly.

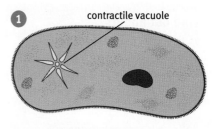
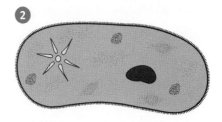
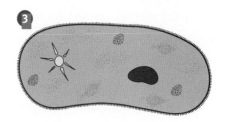
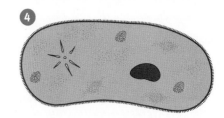

Paramecium is a single-celled organism that lives in fresh water. It has a contractile vacuole that fills and bursts over and over again to get rid of water. If this did not happen, the whole cell would swell and burst.

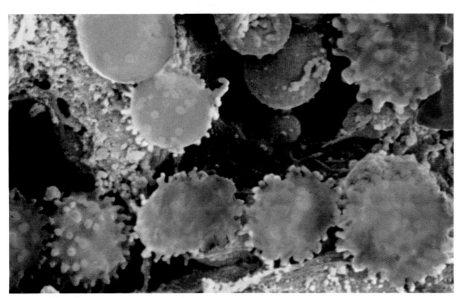

These red blood cells appear wrinkled because they have lost water by osmosis.

About two-thirds of your body is water – in cells, tissue fluid, and blood. Tissue fluid and blood must have a steady water level, or your cells may gain or lose water by osmosis. Your kidneys keep water levels in the body balanced.

Questions

10 The red blood cells in the photo on this page are an unusual shape. They have lost water by osmosis. This can happen when the environment of the cell changes. To produce this effect, were there more dissolved molecules in the blood cells or in the surrounding plasma?

11 What would happen if red blood cells were placed in freshwater?

Key words

active transport

Find out about:

▶ how your body gains and loses water

▶ how your kidneys get rid of wastes

▶ how kidneys balance your water level

F Water homeostasis

Experiments carried out with students showed that being able to drink water in the classroom
▶ increased their concentration time
▶ improved test results

Keeping a steady water level is done by balancing your body's water inputs and water outputs. The diagram below shows how you gain and lose water.

Inputs

water content in:
• food
• drink

water made in:
• respiration

Outputs

water content in:
• exhaled air
• sweat
• urine
• faeces

Water intake and water loss must balance for your body to work well.

Your **kidneys** control the water balance in your body. They do this by changing the amount of urine that you make. On a hot day, or when you have been running, you lose a lot of water in sweat. So your kidneys make a smaller volume of urine. Your urine will be more concentrated that day.

What else do your kidneys do?

Your kidneys have two jobs: water homeostasis and **excretion**. Excretion is getting rid of toxic waste products from chemical reactions in your cells. These two jobs are linked because you use water to flush out waste products such as **urea**.

Liver cells make urea when they break down amino acids your body cannot use. Urea diffuses into your blood and is carried around your body. Except in very low concentrations, urea is poisonous. So as blood passes through your kidneys, urea is filtered out.

How do your kidneys work?

Kidneys work like sieves. Small molecules are filtered out of the blood as it passes through your kidneys. These small molecules are water, sugar (glucose), and urea, and ions of salt. Blood cells and large molecules, such as proteins, are too big, so they stay in the blood.

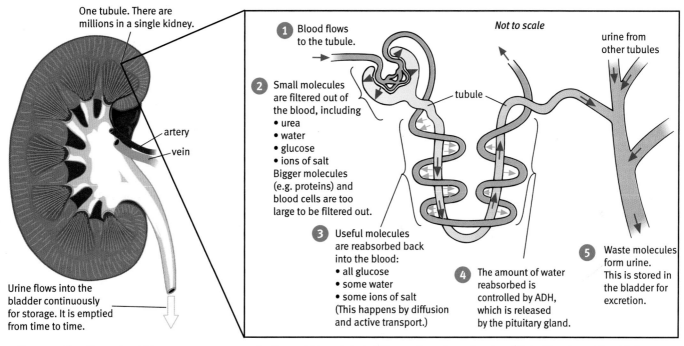

One tubule. There are millions in a single kidney.

artery

vein

Urine flows into the bladder continuously for storage. It is emptied from time to time.

1 Blood flows to the tubule.

Not to scale

urine from other tubules

2 Small molecules are filtered out of the blood, including
• urea
• water
• glucose
• ions of salt
Bigger molecules (e.g. proteins) and blood cells are too large to be filtered out.

tubule

3 Useful molecules are reabsorbed back into the blood:
• all glucose
• some water
• some ions of salt
(This happens by diffusion and active transport.)

4 The amount of water reabsorbed is controlled by ADH, which is released by the pituitary gland.

5 Waste molecules form urine. This is stored in the bladder for excretion.

a Cross-section through a kidney

b How a kidney tubule filters out waste molecules. (*Note:* You do not have to remember the structure of the tubule.)

Getting the balance

Some of the small molecules are useful to the body. You do not want to lose them. So the kidneys **reabsorb** what the body needs. These useful chemicals go back into the blood:

▶ all of the sugar (glucose) for respiration

▶ as much salt as the body needs

▶ as much water as the body needs

The rest of the filtered chemicals go to your bladder. They make up urine.

Key words

kidneys
excretion
urea
reabsorb
urine

Questions

1 List three ways in which your body gains water and three ways in which it loses water.

2 Suggest *two* different things that could cause you to make:

   a a small amount of concentrated urine

   b lots of dilute urine

3 Which molecules do kidneys:

   a filter out of your blood?

   b reabsorb into your blood?

4 Your body is two-thirds water. If your body mass were 60 kg, how many kilogrammes of that would be water?

5 Explain why blood cells or protein in a person's urine is a sign of kidney damage.

More about water balance

Remember that the concentration and volume of your urine varies. On cold days you probably make lots of pale-coloured urine. On hot days you make a smaller volume of darker, more concentrated urine.

The salt concentration of your blood determines how much water your kidneys reabsorb, and how much you excrete in urine. The salt concentration of your blood can become higher than normal because of:

- excess sweating
- not drinking enough water
- eating salty food

Drugs and urine

Some drugs affect the amount of urine a person makes. **Caffeine** in tea and coffee causes a greater volume of dilute urine to be produced. **Alcohol** has an even greater effect and can make people very dehydrated.

The drug **Ecstasy** has the opposite effect. It reduces the volume of urine a person makes. Overheating may also lead to the person drinking too much water. The amount of water in the body can become dangerously high.

Controlling water balance

The control system for water balance is a negative feedback system:

- Receptors in the hypothalamus in your brain detect any changes in salt concentration in the blood.

- The hypothalamus is also the processing centre. When the salt concentration is too high, it triggers the release of a hormone called **ADH** from the **pituitary gland**. This gland is also in the brain, just below the hypothalamus. When the salt concentration is low, no ADH is released.

- The ADH travels in the blood to the kidney tubules. These are the effectors. ADH affects the amount of water that can be reabsorbed back into the blood. The more ADH, the more water is reabsorbed.

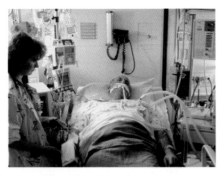

After serious illness or surgery, homeostatic systems in the body may not work reliably. Doctors and nurses monitor patients carefully to keep their bodies in the ideal state to encourage healing and recovery. They may use intravenous drips to prevent dehydration.

Dry air in aeroplanes can cause dehydration. Drinking alcohol and coffee can make the dehydration worse.

Key words

caffeine
alcohol
Ecstasy
ADH
pituitary gland

Questions

6 Describe the affect that these drugs have on urine production:

a caffeine

b alcohol

c Ecstasy

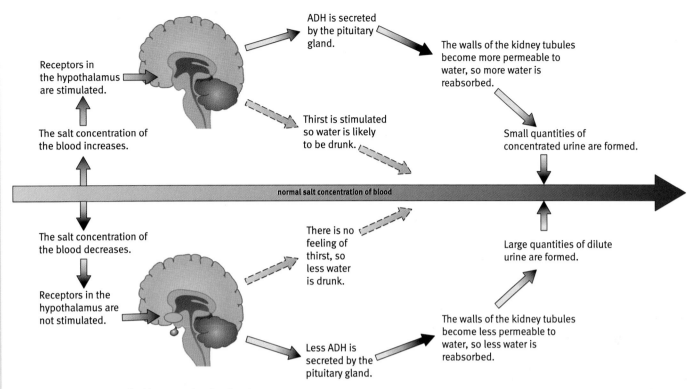

ADH is secreted by the pituitary gland.

Receptors in the hypothalamus are stimulated.

The walls of the kidney tubules become more permeable to water, so more water is reabsorbed.

The salt concentration of the blood increases.

Thirst is stimulated so water is likely to be drunk.

Small quantities of concentrated urine are formed.

normal salt concentration of blood

The salt concentration of the blood decreases.

There is no feeling of thirst, so less water is drunk.

Large quantities of dilute urine are formed.

Receptors in the hypothalamus are not stimulated.

Less ADH is secreted by the pituitary gland.

The walls of the kidney tubules become less permeable to water, so less water is reabsorbed.

Water balance is controlled by negative feedback.

Drugs affect ADH control

Caffeine, alcohol, and Ecstasy change the volume of urine a person makes because they affect ADH production. For example, alcohol suppresses ADH production. Less water is reabsorbed in the kidneys, so a larger volume of urine is made.

Questions

7 ADH is a hormone.

  a Where is ADH made?

  b What affect does ADH have on the body?

8 Copy and complete the table below to show what happens when the salt concentration of the blood changes.

9 Ecstasy triggers release of ADH. Explain the effect this will have on water balance.

10 In a few people the cells that produce ADH are destroyed. Suggest what happens to them and what treatment they need.

	Salt concentration of blood falls	Salt concentration of blood rises
Pituitary gland secretes	less ADH	
Kidney tubules reabsorb		
Urine volume		
Urine concentration		increases

Find out about:
▶ how extreme conditions can upset homeostasis

G When it all goes wrong

Sometimes a person's control systems cannot cope with the conditions. For example, strenuous exercise, very hot or cold climates, or sports like scuba-diving and mountain climbing can all affect homeostasis. In extreme conditions, the person's life is at risk.

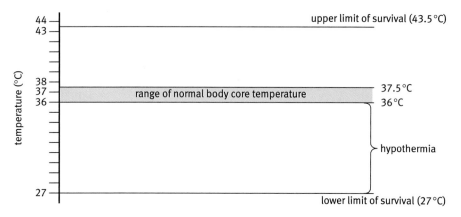

Survival limits for body temperature

Heat stroke

Heat stroke happens when your body cannot lose heat fast enough. For example, in humid conditions, you sweat, but the sweat cannot evaporate. It may be that you have become dehydrated and produce too little sweat. Fever, prolonged exercise, and over-exposure to the sun and drugs such as Ecstasy can all lead to heat stroke.

Core body temperatures of 42 °C and over affect the temperature processing centre in your brain (the hypothalamus). Your temperature control system fails, and you stop sweating. Your body temperature rises out of control.

Ecstasy can cause dehydration because it increases sweat production. Dehydration can lead to heat stroke. (*Note:* Because it affects water control by your kidneys, Ecstasy can also cause over-hydration.)

Symptom of heat stroke	Cause
hot, dry skin	sweating stops
rapid pulse rate	dehydration, stress, increased metabolic rate
dizziness and confusion	nerve cell damage in the brain

Treatment for heat stroke

Rapid cooling of the patient is essential, for example by:

▶ sponging them with water

▶ wrapping them in wet towels

▶ use of a fan

▶ putting ice in their armpits and groin

Doctors may use cooled intravenous drips.

Hypothermia

Hypothermia happens when your core body temperature falls below 35 °C.
It can happen to anyone exposed to low temperatures for long enough.
A person's body heat cannot be replaced as fast as it is being lost. Young
babies and elderly people are at greatest risk, because their temperature
control systems work least well. Babies have a large surface area compared
with their volume, so they lose heat particularly quickly.

Hypothermia causes about 30 000 deaths a year in the UK.

Core body temperature	What happens
below 35 °C	shivering, confusion, drowsiness, slurred speech, loss of coordination
30 °C	coma
28 °C	breathing stops

Climbers are at risk of hypothermia
in freezing external temperatures.
They watch out for signs of
hypothermia in each other.

Treatment for hypothermia

Increasing the patient's core temperature is essential. However, it is
important not to heat their skin and limbs as this increases blood flow to
them. Heat loss then increases, and core temperature falls even further.
This sudden cooling increases the risk of heart failure.

Do

  ▶ insulate them, particularly their head, neck, armpits, and groin

  ▶ handle them gently to keep blood flow to the limbs low

  ▶ warm them gently with warm towels

  ▶ give them warm drinks (not alcohol)

Do not

  ▶ give them food, as digesting food lowers metabolic rate

  ▶ use hot water bottles

Key words

heat stroke
hypothermia

Questions

1 For heat stroke *and* hypothermia list:

  a causes

  b symptoms

  c treatments

2 Explain why a person with heat stroke stops
sweating, even though they need to lose body heat.

3 Why are babies most at risk of hypothermia?

4 Why do people not recognize signs of
hypothermia in themselves?

Summary

In this module you have found out how your body temperature and water levels are kept at a steady state. Homeostasis is vital for cells to be able to work properly.

Keeping a steady state

▶ Automatic control systems in the body keep factors steady, including body temperature and water level.

▶ Artificial control systems are similar to body control systems; they both have receptors, processing centres, and effectors.

▶ Strenuous exercise and extreme climates can affect homeostasis (body temperature, water balance, blood oxygen level H, salt level H).

▶ Negative feedback is used in many control systems, so that any change from a normal level causes a response that returns the system back to normal. H

▶ Some effectors work antagonistically, which gives a more sensitive response. H

Moving in and out of cells

▶ Chemicals move in and out of cells by diffusion, e.g. O_2, CO_2, dissolved food.

▶ Diffusion is the passive movement of molecules from where they are in high concentration to where they are in low concentration.

▶ Osmosis is the overall movement of water molecules from a dilute to more concentrated solution through a partially permeable membrane.

▶ Keeping the body's water level balanced is important so that cell contents are at the correct concentration for the cell to work properly.

▶ Some chemicals, e.g. glucose, are moved in or out of cells by active transport, which requires energy. H

Enzymes

▶ Enzymes are proteins that speed up chemical reactions in cells.

▶ A small part of an enzyme (the active site) is shaped so that only molecules with the correct shape can fit with the enzyme; this is the 'lock-and-key' model.

▶ The 'lock-and-key' model explains why each type of enzyme only speeds up one particular chemical reaction.

▶ Enzymes work best at a certain temperature, which is why the body's temperature must be kept at a steady level.

▶ At low temperatures, increasing the temperature will increase the rate of a reaction because molecules collide more often and with more energy.

▶ At high temperatures an enzyme stops working (it is denatured) because its active site shape is changed.

Body temperature

▶ Energy gain and loss must be balanced to keep a steady body temperature.

▶ Receptors in the skin and brain detect changes in temperature of the air and the blood.

▶ The brain processes information about temperature control in the hypothalamus, and effectors (sweat glands and muscle) produces responses to keep body temperature steady.

▶ Vasodilation and vasoconstriction control the amount of blood flowing near to the skin surface. H

▶ You should be able to describe the cause, symptoms, and first aid treatment for heatstroke and hypothermia.

Body water balance

▶ Water is gained from food, drinks, and respiration, and is lost through sweating, breathing, faeces, and urine.

▶ The amount and concentration of urine made by the body depend on several factors.

▶ Kidneys remove waste urea from the body and balance the level of other chemicals in the blood.

▶ Concentration of urine is controlled by the hormone ADH, released by the pituitary gland in the brain. ADH production is controlled by negative feedback.

▶ Alcohol and Ecstasy both effect the amount and concentration of urine made by the body (by affecting ADH production H).

Questions

1 **a** Write definitions for these key words:

    i homeostasis **ii** stimulus **iii** response
    iv receptor **v** effector

  b Explain why it is important that these factors are kept at steady levels in the body:

    i temperature   **ii** water

  c Draw a flowchart to explain how the body reacts when body temperature increases above normal.

2 A person sprays air freshener at the other side of a room. After a few minutes you can smell the air freshener molecules.

  a Name the process by which the air freshener molecules have moved towards you.

  b Write down a definition to describe this process.

  c Say whether this process is passive (no energy needed) or active (requires energy).

  d Name two chemicals that move in and out of body cells by this process.

3 Water moves in and out of cells.

  a Name the process by which this happens.

  b Write bullet point notes to explain why a single-celled organism in fresh water will gain water. Start with:

    • The solution around the cell is more dilute than the cell contents.

4 **a** Draw a graph showing how temperature effects the rate of an enzyme reaction.

  b Label your graph with these notes:

    A: At low temperatures, increasing the temperature increases the frequency and energy of collisions. The rate of reaction increases.

    B: The enzyme has an optimum temperature. It works at its fastest rate.

    C: At higher temperature, the active site changes shape. The enzyme is denatured and stops working.

  c The enzyme catalase speeds up the breakdown of hydrogen peroxide to water and oxygen. Explain why catalase cannot speed up any other reactions. Use the 'lock-and-key' model in your answer.

5 Some chemicals are filtered out of the blood when it goes through a kidney. The table below shows the concentration of some chemicals at different places.

A: blood entering the kidney

B: liquid just after it is filtered from the blood

C: filtered liquid just before it leaves the kidney

  a Name one chemical that is not filtered out by the kidney.

Where	Concentration (g in 100 cm³)		
	Protein	Glucose	Urea
A	7.5	0.1	0.03
B	0.0	0.1	0.03
C	0.0	0.0	0.15

  b Explain why this is not filtered out.

  c What happens to the glucose between **B** and **C**?

  d Why does the concentration of urea increase between **B** and **C**?

6 Alcohol causes more dilute urine to be produced.

  a Explain why this can lead to dehydration.

  b Describe how alcohol interferes with ADH control of water balance. H

C4

Why study chemical patterns?

The periodic table is so important in chemistry because it helps to make sense of the mass of information about all the elements and their compounds. The table offers a framework that can give meaning to all the facts about properties and reactions.

The science

What fascinates chemists is that it is possible to use ideas about atomic structure to explain the periodic table and the properties of different elements.

Light and electrons in atoms can affect each other. This is the science behind spectroscopy. At first, spectroscopy led to the discovery of new elements. Today, a wide range of techniques means that spectroscopy provides the essential tools for studying chemicals and chemical reactions.

Chemistry in action

The ideas in this module are of great practical importance. Many of the most sensitive methods of chemical analysis depend on spectroscopy.

Without ionic theory there would be no aluminium metal. Ionic theory is also vital in explaining how our nerves and brain work.

Chemical patterns

Find out about:

▶ the chemistry of some very reactive elements

▶ the patterns in the periodic table

▶ how scientists can learn about the insides of atoms

▶ the use of atomic theory to explain the properties of chemicals

▶ the ways in which atoms become charged and turn into ions

Find out about:
▶ relative masses of atoms
▶ periodic patterns
▶ groups and periods

A The periodic table

A century of discovery

There were only about 30 known elements when Horatio Nelson led the British fleet to victory at Trafalgar in 1805. Nearly a hundred years later, when Queen Victoria died, scientists had discovered all but three of the stable elements found on Earth.

The discovery of so many elements encouraged chemists to look for patterns in the properties of the elements. One idea, which seemed strange at first, was to look for a connection between the chemistry of elements and the masses of their atoms.

Relative atomic masses

In the 1800s, scientists could not measure the actual masses of atoms – they could only compare them. They chose to compare the masses of atoms with the mass of the lightest atom, hydrogen. On the **relative atomic mass** scale the relative mass of a hydrogen atom is 1, that of a carbon atom is 12, that of an oxygen atom is 16, and so on.

Germanium – one of Mendeléev's missing elements. He used his version of the periodic table to predict that the missing element would be a grey metal that would form a white oxide with a high melting point. He also predicted that its chloride would boil below 100 °C and have a density of about 1.9 g/cm³.

Johann Döbereiner, a German scientist, noticed that there were several examples of groups of three elements with similar properties (for example calcium, strontium, and barium). For each group, the relative mass of the atoms of the middle element was the mean of the relative masses of the other two elements.

This, and other early attempts to find connections between chemical properties and atomic mass, were not taken seriously at the time.

Elements in order

Dmitri Mendeléev, a Russian scientist, showed that it is possible to come up with patterns with real meaning when elements are lined up in order of the masses of their atoms (their relative atomic masses). Mendeléev's inspiration was to realize that not all of the elements had yet been discovered. He left gaps for missing elements when this was necessary to produce a sensible pattern.

When Mendeléev put the elements in order of relative atomic mass, he spotted that at intervals along the line there were elements with similar properties. Using elements known today, for example, you can see that the third, eleventh, and nineteenth elements (lithium, sodium, and potassium) are very similar.

Periodicity

Gold, platinum, titanium, and other transition metals are used to make jewellery. They are shiny metals which do not react with the air.

A repeating pattern of any kind is a **periodic** pattern. An example is the repeating pattern on a roll of fabric or wallpaper. The table of the elements gets its name from the repeating, or periodic, patterns you see when chemists line up the elements in order.

The periodic table now

In the periodic table, the elements are arranged in rows, one above the other. Each row is a **period**. The most obvious repeating pattern is from metals on the left to non-metals on the right. Every period starts with a very reactive metal in group 1 and ends with an unreactive gas in group 8.

Elements with similar properties fall into a column. Each column is a **group** of similar elements.

Key words
relative atomic mass
periodic
period
group

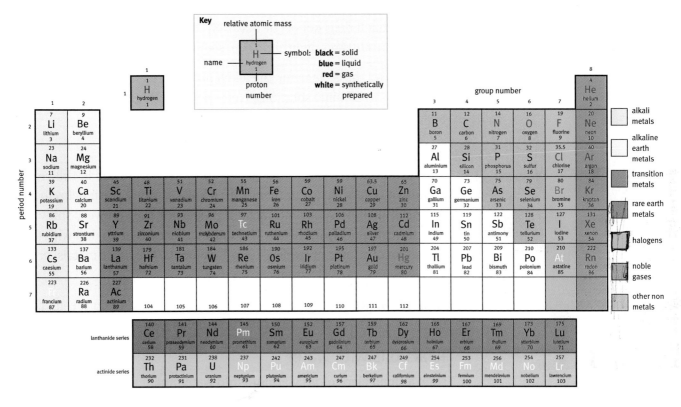

The periodic table. Over three-quarters of the elements are metals. They lie to the left of the table. Most of them are in groups 1 and 2 and the block of transition metals.

Questions

1 In the periodic table identify and name:
   a a liquid halogen
   b an alkali metal that does not occur naturally
   c a gaseous element with properties similar to sulfur
   d a solid element similar to chlorine
   e a liquid metal with properties similar to zinc

2 How many times heavier than a hydrogen atom are the atoms of:
   a carbon?
   b magnesium?
   c bromine?

3 How many times heavier is:
   a a magnesium atom than a carbon atom?
   b a sulfur atom than a helium atom?
   c an iron atom than a nitrogen atom?

4 Name two elements in the modern periodic table that break Mendeléev's rule that the elements should be arranged in order of relative atomic mass.

5 Explain how Mendeléev could predict the properties of the unknown element germanium from what was known about other elements such as silicon and tin.

Find out about:
- group 1 metals
- reactions with water and chlorine
- similarities and differences between group 1 elements

B The alkali metals

The metals in group 1 of the periodic table are very reactive. They are so reactive that they have to be kept under oil to stop them reacting with oxygen or moisture in the air.

There are six elements in the group. Two of them, rubidium and caesium, are so reactive and rare that you are unlikely to see anything of them except on video. A third, francium, is highly radioactive. Its atoms are so unstable that it does not occur naturally. As a result, the study of group 1 usually concentrates on lithium (Li), sodium (Na), and potassium (K).

Chemists call these elements the alkali metals because they react with water to form alkaline solutions. Note that it is the compounds of these metals which are alkalis and not the metals themselves.

Strange metals

Most metals are hard and strong. The alkali metals are odd in this respect because it is possible to cut them with a knife. Cutting them helps to show up one of their most obvious metallic properties: they are very shiny but they tarnish quickly in the air. The shiny surface becomes dull with the formation of a layer of oxide. Group 1 elements, like other metals, are also good conductors of electricity.

Most metals are dense and have high melting points. Again the alkali metals are odd: they float on water and melt on very gentle heating.

Reactions with water

Drop a small piece of grey lithium into water and it floats, fizzes gently, and disappears as it turns into lithium hydroxide (LiOH). It dissolves, making the solution alkaline. It is possible to collect the gas and use a burning splint to show that it is hydrogen.

lithium + water → lithium hydroxide + hydrogen

The reaction with sodium is more exciting. The reaction gives out enough energy to melt the sodium, which skates around on the surface of the water. Sometimes sparks from the molten sodium ignite the hydrogen formed, which then burns with a yellowish flame. Like lithium, the sodium turns into its hydroxide (NaOH) and dissolves to give an alkaline solution.

The reaction with potassium is very violent. The hydrogen given off catches fire at once, and molten metal may be thrown from the surface of the water. The result is an alkaline solution of potassium hydroxide (KOH).

Cutting a lump of sodium to show a fresh surface of the metal

Pellets of the alkali sodium hydroxide. The traditional name is caustic soda. Anything caustic attacks skin. Alkalis such as NaOH are more damaging to skin and eyes than many acids.

Reactions with chlorine

Hot sodium burns with a bright yellow flame. It produces clouds of white sodium chloride crystals (NaCl). This is everyday 'salt', used for seasoning food.

The other alkali metals react in a similar way with chlorine. Lithium produces lithium chloride (LiCl). Potassium produces potassium chloride (KCl). Like everyday salt, these compounds are also colourless, crystalline solids which dissolve in water.

Chemists use the term **salt** to cover all the compounds of metals with non-metals. So the chlorides of lithium, sodium, and potassium are all salts.

Trends

The alkali metals are all very similar, but they are not identical. There are clear **trends** in their properties down the group from lithium to sodium to potassium. These trends cover both **physical properties**, such as density and melting point, and **chemical properties**, such as the reactivity of the metals with water and chlorine.

Compounds of the alkali metals

The compounds of the alkali metals are very different from the elements. The elements are dangerously reactive. But chlorides of sodium and potassium, for example, have a vital role to play in the blood and in the way in which our nerves work.

Many compounds of alkali metals are soluble in water. Soluble sodium compounds, in particular, make up a number of common everyday chemicals. All homes contain sodium chloride ('salt'). Other important domestic products include sodium hydroxide (in oven cleaners), sodium hypochlorite (in bleach), and sodium hydrogencarbonate (as the bicarbonate of soda in antacids).

Sodium burning in chlorine gas

Key words

salt
trends
physical properties
chemical properties

Questions

1 Arrange the names of the alkali metals Li, Na, and K in order of reactivity with water, placing the most reactive of the metals first.

2 Predict these properties of rubidium:
   a How easily can it be cut with a knife?
   b What happens to a fresh-cut surface of the metal in the air?
   c What happens if you drop a small piece of rubidium onto water?

3 For the hydroxide of rubidium predict:
   a its colour
   b its formula
   c whether or not it is soluble in water

4 For the chloride of caesium predict:
   a its colour   b its formula
   c whether or not it is soluble in water

5 Give an example to show that the trend is for the alkali metals to become more chemically reactive down the group from lithium to potassium.

6 Suggest and explain the precautions necessary when demonstrating the reaction of potassium with water.

Find out about:
▶ chemical symbols
▶ formulae
▶ balanced equations

c Chemical equations

Equations are important because they do for chemists what recipes do for cooks. They allow chemists to work out how much of the starting materials to mix together and how much of the products they will then get.

Chemical models

In a **chemical change**, there is no change in mass because the number of each type of atom stays the same. The atoms regroup but no new ones appear and none are destroyed during a chemical reaction.

Hydrogen burns in oxygen to form **molecules** of water. The models in the figure below show what happens. In each water molecule there is only one oxygen atom. So one oxygen molecule reacts with two molecules of hydrogen to make two water molecules. There are equal numbers of hydrogen atoms and oxygen atoms on each side of the arrow. The term **chemical equation** can be used.

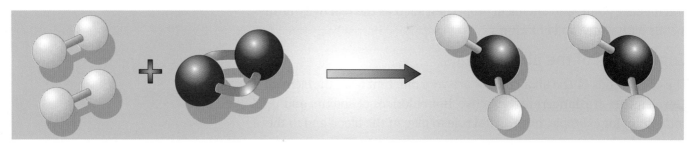

A model equation for the reaction of hydrogen with oxygen

Chemical symbols

Using drawings or photographs of models to describe every reaction would be very tiresome. Instead chemists write symbol equations to show the numbers and arrangements of the atoms in the reactants and the products.

When written in symbols the equation in the figure above becomes

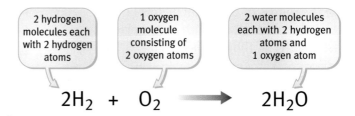

| 2 hydrogen molecules each with 2 hydrogen atoms | 1 oxygen molecule consisting of 2 oxygen atoms | 2 water molecules each with 2 hydrogen atoms and 1 oxygen atom |

$$2H_2 \ + \ O_2 \longrightarrow 2H_2O$$

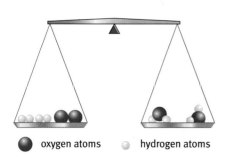

● oxygen atoms ○ hydrogen atoms

The reactants and the products of a reaction 'balance'

The equation is 'balanced' because it has the same number of atoms of each type on the left (of the arrow) and on the right. It balances in the literal sense too. The reactants have the same mass as the products, so the chemicals on the right balance the chemicals on the left, if placed on an old-fashioned pair of scales.

Formulae

You cannot write an equation unless you first know

▶ all the starting chemicals (the reactants)

▶ everything that is formed during the change (the products)

When writing an equation, you have to write down the correct chemical **formulae** for the reactants and products. Chemists have worked these out by experiment, and you can look them up in data tables.

If the element or compound is molecular, you write the formula for the molecule in the equation. This applies to most non-metals (O_2, H_2, Cl_2) and most compounds of non-metals with non-metals (H_2O, HCl, NH_3).

Not all elements and compounds consist of molecules. For all metals, and for the few non-metals that are not molecular (C, Si), you just write the symbol for a single atom.

The compounds of metals with non-metals are also not molecular. For these compounds you write the simplest formula for the compound, such as $LiOH$, $NaCl$, potassium carbonate (K_2CO_3).

Writing balanced equations

Follow the four steps shown in the margin to write **balanced equations**.

Example

Write down a balanced equation to show the reaction of natural gas (methane, CH_4) with oxygen.

Step 1 Describe the reaction in words:

methane + oxygen → carbon dioxide + water

Step 2 Write down the formulae for the reactants and products:

$$CH_4 + O_2 → CO_2 + H_2O$$

Step 3 Balance the equation:

You must not change any of the formulae. You balance the equation by writing numbers in front of the formulae. These numbers then refer to the whole formula.

$$CH_4 + 2O_2 → CO_2 + 2H_2O$$

Step 4 Add state symbols:

State symbols usually show the states of the elements and compounds at room temperature and pressure. The chemicals in an equation may be solid (s), liquid (l), gaseous (g), or dissolved in water (aq, for aqueous).

$$CH_4(g) + 2O_2(g) → CO_2(g) + 2H_2O(l)$$

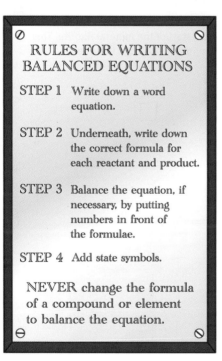

RULES FOR WRITING BALANCED EQUATIONS

STEP 1   Write down a word equation.

STEP 2   Underneath, write down the correct formula for each reactant and product.

STEP 3   Balance the equation, if necessary, by putting numbers in front of the formulae.

STEP 4   Add state symbols.

NEVER change the formula of a compound or element to balance the equation.

Questions

1 Write balanced symbol equations for these reactions of the alkali metals:

  a sodium with water

  b potassium with water

  c sodium with chlorine

  d lithium with bromine

  e potassium with iodine

Find out about:

▶ group 7 elements
▶ halogen molecules
▶ similarities and differences between group 7 elements

Crystals of the mineral fluorite. This mineral is calcium fluoride.

D The halogens

Salt formers

Fluorine, chlorine, bromine, and iodine are all very reactive non-metals. They are interesting because of their vigorous chemistry. As elements they are hazardous because they are so reactive. For the same reason, they are not found free in nature. They occur as compounds with metals.

It is not normally possible to study fluorine because it is so dangerously reactive.

The name 'halo-gen' means 'salt-former'. These elements form salts when they combine with metals. Examples include everyday 'salt' itself, which occurs as the minerals halite (NaCl) and fluorite (CaF_2) found as the mineral Blue John (used in jewellery) in Derbyshire caves.

Non-metal patterns

Non-metals typically have low melting and boiling points. Chlorine is a greenish gas at room temperature. Bromine is a dark-red liquid which easily turns to an orange vapour. Iodine is a dark-grey solid which turns to a purple vapour on gentle warming.

The **halogens**, like most non-metals, are molecular. They each consist of molecules with the atoms joined in pairs: Cl_2, Br_2, and I_2. The forces between the molecules are weak, and so it is easy to separate them and turn the halogens into gases.

- dense, pale-green gas
- smelly and poisonous
- occurs as chlorides, especially sodium chloride in the sea

- deep red liquid with red–brown vapour
- smelly and poisonous
- occurs as bromides, especially magnesium bromide in the sea

- grey solid with purple vapour
- smelly and poisonous
- occurs as iodides and iodates in some rocks and in seaweed

Chlorine, bromine, and iodine

Halogen patterns

All the halogens can harm living things. They can all kill bacteria. Domestic **bleach** is a solution of chlorine in sodium hydroxide sold to disinfect worktops and toilets. In the days before modern antiseptics, people used a solution of iodine to prevent infection of wounds.

The bleaching effect of the halogens illustrates the general trend in reactivity down the group. The usual laboratory test for chlorine shows that the gas quickly bleaches moist indicator paper. Bromine vapour also bleaches vegetable dyes such as litmus, but more slowly. Iodine has a slight bleaching effect too, but it also stains paper brown, which masks the change.

The reaction with iron also illustrates the clear trend in the reactivity of the elements down the group. Hot iron glows brightly in chlorine gas. The product is iron chloride (FeCl3), which appears as a rust-brown solid. Iron also glows when heated in bromine vapour, but less brightly. There is even less sign of reaction when iron is heated in iodine vapour.

Practical importance

While the halogens themselves are too hazardous for everyday use, their compounds are of great practical importance.

The chemical industry turns everyday salt (NaCl) into chlorine (and sodium) and then uses the chlorine to make plastics such as polyvinylchloride (PVC). Another large-scale use of chlorine is water treatment to stop the spread of diseases. So chlorine compounds offer many benefits, but there are hazards too. Some chlorine compounds are so stable that they persist in the natural environment, where they can be a threat to life or bring about long-term damage. A dramatic example is the impact of chlorofluorocarbons (CFCs) and other halogen compounds on the concentration of ozone in the upper atmosphere.

Most of the bromine we use comes from the sea. Liquid bromine itself is extremely corrosive. However, the chemical industry makes important bromine compounds. These include medical drugs, and pesticides to protect food crops.

Traces of iodine compounds are essential to a healthy human diet. In regions where there is little or no natural iodine, it is usual to add potassium iodide to everyday table salt or to drinking water. This prevents disease of the thyroid, a gland in the neck. Iodine and its compounds are starting materials for the manufacture of medicines, photographic chemicals, and dyes.

Hot iron in a jar of chlorine

Key words

halogens
toxic
corrosive
harmful
bleach

Questions

1 Which halogen is dangerously corrosive?

2 Write balanced equations for the reactions of:
   a iron with chlorine to form FeCl$_3$
   b potassium with chlorine
   c lithium with iodine

3 Fluorine (F$_2$) is the first element in group 7. Predict the effect of passing a stream of fluorine over iron. Write an equation for the reaction.

Robert Bunsen (1811–1899), who discovered the flame colours of elements with the help of his new burner

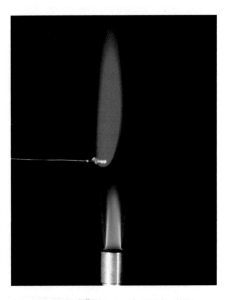

The bright red flame produced by lithium compounds. The compounds of other elements also produce colours in a flame:
• sodium – bright yellow
• potassium – lilac
• calcium – orange–red
• barium – green

E The discovery of helium

A new burner for chemistry

Robert Bunsen moved to the University of Heidelberg in Germany in 1852. Before taking up the job as professor of chemistry, he insisted on having new laboratories. He also demanded gas piping to bring fuel from the gas works – this had just opened to light the city streets.

Existing burners produced smoky and yellow flames. Bunsen wanted something better. In 1855 he invented the type of burner that is still used today in laboratories all over the world.

The great advantage of Bunsen's burner was that it could be adjusted to give an almost invisible flame. Bunsen used his burner to blow glass. He noticedthat whenever he held a glass tube in a colourless flame, the flame turned yellow.

Flame colours

Soon Bunsen was experimenting with different chemicals, which he held in the flame at the end of a platinum wire. He found that different chemicals produced characteristic **flame colours**.

Bunsen thought that this might lead to a new method of chemical analysis, but he soon realized that it seemed only to work for pure compounds. It was hard to make any sense of flames from mixtures. So he mentioned his problem to Gustav Kirchhoff, who was the professor of physics.

Flame spectra

'My advice as a physicist', said Kirchhoff, 'is to look not at the colour of the flames, but at their spectra.'

Kirchhoff built a spectroscope, by putting a glass prism into a wooden box and inserting two telescopes at an angle. Light from a flame entered through one telescope. It was split into a spectrum by the prism and then viewed with the second telescope.

Bunsen and Kirchhoff soon found that each element has its own characteristic spectrum when its light passes through a prism. Each spectrum consists of a set of lines. With their spectroscope they were able to record the **line spectra** of many elements.

The spectrum of cadmium flame is made up of series of lines. Note the difference with the continuous spectrum of white light.

Using **spectroscopy**, Bunsen discovered two new elements in the waters of Durkheim Spa. He based their names on the colours of their spectra. He called them caesium and rubidium from the Latin for 'sky blue' and 'dark red'.

A Sun element

In 1868 there was a total eclipse of the Sun. Normally, the blinding light from the centre of the Sun makes it impossible to see the much fainter light from the hot gases around the edges of the star. During an eclipse, the Moon hides the whole bright disc of the Sun but not the much fainter light from the hot gases around the edges. This makes it possible to study this light from these gases.

Pierre Janssen, a French astronomer, took very careful observations of the Sun's spectrum during the eclipse. In the spectrum of the light, he saw a yellow line where no yellow line was expected to be.

Excited by these observations, both Janssen and an English astronomer, Joseph Lockyer, developed new methods to study the light from the Sun's gases. They worked independently, but both came to the same conclusion. There must be an unknown element in the Sun, producing the unexpected yellow line in the spectrum. Janssen and Lockyer published their findings at almost the same time – a coincidence which led to them becoming good friends.

The new element was called 'helium' from the Greek word *helios*, meaning Sun. Both astronomers were still alive in 1895 when William Ramsay, a British chemist, used spectroscopy to discover helium on Earth. The gas came from boiling up a uranium ore with acid.

A solar eclipse in 1868 helped scientists to discover helium. During an eclipse it is possible to study the spectra of the light from the hot gases around the edges of the Sun.

> **WARNING!**
> Never look directly at the Sun, even during an eclipse. You can damage your eyes or even be blinded!

Questions

1 Why was it important for Bunsen to have a burner with a colourless flame?

2 Why is it not possible to analyse chemical mixtures simply by looking at their flame colours?

3 Use your knowledge of group 1 chemistry to suggest an explanation for the fact that rubidium and caesium were not discovered until the technique of spectroscopy was developed.

4 Suggest reasons why helium was discovered on the Sun before it was discovered on Earth.

5 Why is it rare for two or more scientists who make the same discovery at the same time to end up as friends?

Key words
flame colour
line spectra
spectroscopy

Find out about:
▶ atomic theory
▶ the nuclear model of the atom
▶ protons, neutrons, and electrons

Part of the map of the London Underground

F Atomic structure

Atomic models

A picture, or model, of an atom can be used to understand how atoms join together to form compounds and how they regroup during chemical reactions.

Scientists use different models to solve different problems. There is not one atomic theory that is 'true'. Each model can represent only a part of what we know about atoms.

It is like using maps to travel through London. The usual underground rail map is a very useful guide for getting from one tube station to another. It is 'true' in that it shows how the lines and stations connect, but it cannot solve all of a traveller's problems. The map does not show how the tube stations relate to roads and buildings on the surface. For that you need a street map.

1804
Dalton's solid atom

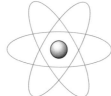

1913
The Bohr–Rutherford 'Solar System' atom, in which electrons orbit round a very small nucleus

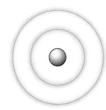

1924
A model of the atom in which the electrons are no longer treated as particles but pictured as occupying energy levels, which give rise to regions of negative charge around the nucleus (charge clouds)

1932
The atom in which the nucleus is built up from neutrons as well as protons

2000+
The present-day atom in which the nucleus is built up from many kinds of particles

Atomic models from 1800 to the present. The diameter of an atom is about ten million times smaller than a millimetre. These diagrams are distorted. On this scale the nuclei would be invisibly small.

Dalton's atomic theory

The story of our modern thinking about atomic structure began with John Dalton at the beginning of the nineteenth century. In Dalton's theory everything is made of atoms that cannot be broken down. The very word 'atom' means 'indivisible'.

The main ideas in Dalton's theory still apply to everyday chemistry. So far as chemistry is concerned, each element does have its own kind of atom, and the atoms of different elements differ in mass. The idea that equations must balance is based on Dalton's view that atoms are not created or destroyed during chemical changes.

Even so, Dalton's theory is limited. It cannot explain the pattern of elements in the periodic table. Nor can it explain how atoms join together in elements and compounds.

Key words
nucleus
protons
neutrons
electrons
proton number

Inside the atom

In time it became clear that atoms are not solid, indivisible spheres. From the middle of the nineteenth century, scientists began to find ways of exploring the insides of atoms. Still today, scientists are spending vast sums of money to build particle accelerators (atom-smashers) which work at higher and higher energies. They hope to discover more about the fine structure of atoms.

It is possible to explain much more about the chemistry of elements and compounds with the help of a model of atomic structure that includes sub-atomic particles.

A model for chemistry

In your study of chemistry you will be using an atomic model which dates back to 1932, when James Chadwick discovered the neutron. In this model the mass of the atom is concentrated in a tiny, central **nucleus**. The nucleus consists of **protons** and **neutrons**. The protons have a positive electric charge. Neutrons are uncharged.

Around the nucleus are the **electrons**. The electrons are negatively charged. The mass of an electron is so small that it can often be ignored. In an atom, the number of electrons equals the number of protons in the nucleus (the **proton number**). This means that the total negative charge equals the total positive charge, and overall an atom is uncharged.

The Large Hadron Collider at CERN in Switzerland is a particle accelerator that will probe more deeply into matter than ever before. Due to switch on in 2007, it will ultimately collide beams of protons with very high energies.

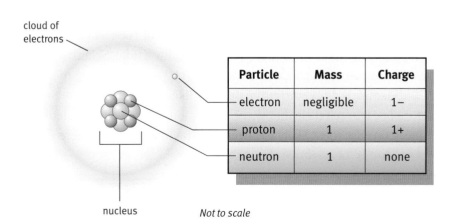

Particle	Mass	Charge
electron	negligible	1−
proton	1	1+
neutron	1	none

cloud of electrons

nucleus

Not to scale

All atoms consist of these three basic particles. The nucleus of an atom is very, very small. The diameter of an atom is about ten million million times greater than the diameter of its nucleus.

Questions

1 With the help of the periodic table on page 39, work out:

 a the element with one more proton in its nucleus than a chlorine atom

 b the element with one proton fewer in its nucleus than a neon atom

 c the number of protons in a sodium atom

 d the number of electrons in a bromine atom

 e the size of the positive charge on the nucleus of a fluorine atom

 f the total negative charge on the electrons in a potassium atom

Find out about:

- evidence for energy levels
- electrons in shells
- electron configurations

G Electrons in atoms

Electrons in orbits

In 1913, the Danish scientist Niels Bohr came up with an explanation for the line spectra from atoms. He made a close study of the spectrum from hydrogen.

In the Bohr model for atoms, the electrons orbit the nucleus as the planets orbit the Sun. Bohr's idea was that heating atoms gives them energy. This forces the electrons to move to higher-energy orbits further from the nucleus. These electrons then drop back from outer orbits to inner orbits. They give out light energy as they do so. Each energy jump corresponds to a particular colour in the spectrum. The bigger the jump, the nearer the line to the blue end of the spectrum. Only certain energy jumps are possible, so the spectrum consists of a series of lines.

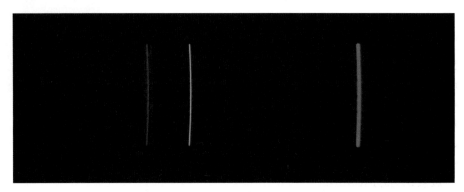

The line spectrum of hydrogen. Atomic theory can explain why this spectrum is a series of lines.

Bohr was able to use his theory to calculate sizes of the energy jumps. He could then deduce the energy levels of electrons in the various orbits.

Electrons in shells

The comparison with the Solar System and the use of the term orbit can be misleading. The theory has moved on since Bohr's time. Scientists still picture the electrons at a particular **energy level**. However, in the modern theory the electrons do not orbit the nucleus like planets round the Sun. All that theory can tell us is that there are regions around the nucleus where electrons are most likely to be found. Chemists describe these regions as 'clouds' of negative charge.

Think of each electron cloud as a **shell** around the nucleus. Each shell is one of the regions in space where there can be electrons. The shells only exist if there are electrons in them. Electrons in the same shell have the same energy.

Key words

energy level
shell
electron configuration

Electron configurations

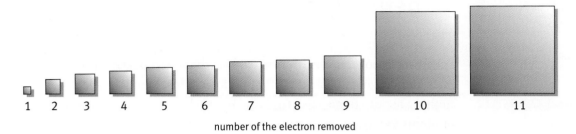

The areas of the squares are in proportion to the amount of energy needed to remove the electrons one by one from a sodium atom.

Each electron shell can contain only a limited number of electrons. The innermost shell with the lowest energy fills first. When full, the electrons go into the next shell. Evidence for this theory comes not only from spectra but also from measurements of the energy needed to remove electrons from atoms.

There are eleven electrons in a sodium atom (proton number 11). Scientists have measured the quantities of energy needed to remove these electrons one by one from a sodium atom. The relative values are represented by the areas of the squares in the picture above. You can see that it is quite easy to remove the first electron. The next eight are more difficult to remove. Finally it becomes really hard to remove the last two electrons, which are held very powerfully because they are in the shell closest to the nucleus.

This supports the idea that the electrons in a sodium atom are arranged in three shells as shown in the figure on the right. The diagram shows common representations of the **electron configuration** of the element.

The first shell that is closest to the nucleus can hold up to two electrons. The second shell can hold eight. Once the second shell holds eight electrons, the third shell starts to fill.

If there are more electrons, they occupy further shells. After the first twenty elements the arrangements become increasingly complex as the shells hold more electrons and the energy differences between shells get smaller.

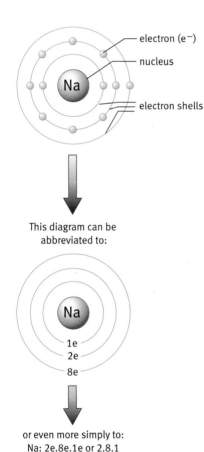

This diagram can be abbreviated to:

or even more simply to:
Na: 2e.8e.1e or 2.8.1

Two-dimensional representations of the electrons in shells in a sodium atom

Questions

1 Draw diagrams to show the electrons in shells for these atoms:
   a beryllium   **b** oxygen   **c** magnesium
   Refer to the periodic table on page 39 for the proton numbers, and therefore the number of electrons, in each atom.

2 How does the diagram at the top of the page support the representations of a sodium atom in the diagram underneath?

Find out about:
- atomic structure and periods
- electron configurations and groups
- explaining similarities and differences between the elements

H Electronic structures and the periodic table

The periodic table then and now

Scientists discovered electrons in 1897, which was nearly thirty years after Mendeléev published his first periodic table. Mendeléev knew nothing about atomic structure and he used the relative masses of atoms to put the elements in order.

A modern periodic table shows the elements in order of proton number, which is also the number of electrons in an atom. One of the convincing pieces of evidence for the 'shell model' of atomic structure is that it can help to explain the patterns in the periodic table.

Periods

The diagram below shows the connection between the horizontal rows of the periodic table and the structure of atoms. From one atom to the next, the proton number increases by one and the number of electrons increases by one. So the electron shells fill up progressively from one atom to the next.

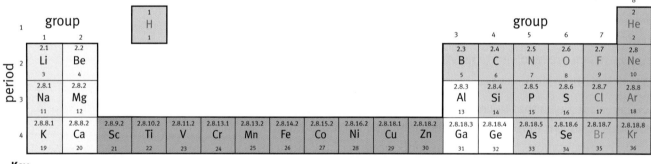

Key

2.4 — number of electrons in each shell
C — symbol
6 — proton number

Electron configurations for the first 20 elements in the periodic table

The first period from hydrogen to helium corresponds to filling the first shell. The second shell fills across the second period from lithium (2.1) to neon (2.8). Eight electrons go into the third shell from sodium (2.8.1) to argon (2.8.8) and then the third shell starts to fill from potassium to calcium.

In fact the third shell can hold up to 18 electrons. This shell is completed from scandium to zinc, before the fourth shell continues to fill from gallium to krypton. This accounts for the appearance of the block of transition metals in the middle of the table. Why this happens cannot be explained by the simple theory described here. You will find out the explanation if you go on to a more advanced chemistry course.

⊞ Groups

When atoms react, it is the electrons in their outer shells which get involved as chemical bonds break and new chemicals form. It turns out that elements have similar properties if they have the same number and arrangement of electrons in the outer shells of their atoms.

Three of the alkali metals appear in the diagram on the right. You can see that they each have one electron in the outer shell of their atoms. This is the case for the other alkali metals too. This helps to account for the similarities in the chemistries of these elements.

The alkali metals are not all the same because their atoms differ in the number of inner full shells. A sodium atom has two inner filled shells, so it is larger than a lithium atom, and its outer electron is further away from the nucleus. As a result, the two metals have similar, but not identical, physical and chemical properties.

Metals and non-metals

Elements with only one or two electrons in the outer shell are metals. Elements with more electrons in the outer shell are generally non-metals, though there are exceptions to this, such as aluminium, tin, and lead. The halogens are non-metal elements with seven electrons in the outer shell.

At the end of each period there is a noble gas. This is a group of very unreactive elements. The first member of the group is helium.

The term 'noble' has been used by alchemists and chemists for hundreds of years to describe elements that are inert to most common reagents. The chemical nobility stand apart from the hurly-burly of everyday reactions.

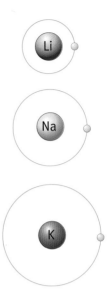

The trend in the size of the atoms of group 1 elements reflects the increasing number of full, inner electron shells down the group. Only the outer shells are shown here.

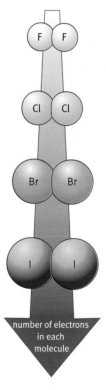

The trend in the size of the molecules of group 7 elements reflects the increasing number of full, inner electron shells in the atoms down the group.

Questions

1 Explain the meaning of this statement: the electron configuration of chlorine is (2.8.7).

2 a What are the electron configurations of the elements beryllium, magnesium, and calcium?

 b In which group of the periodic table do these three elements appear?

 c Are the elements metals or non-metals?

Find out about:
- salts
- properties of salts
- electricity and salts

Salts

Why are salts so different from their elements?

Compounds of metals with non-metals are salts. Chemists can explain the differences between a salt and its elements by studying what happens to the atoms and molecules as they react. A good example is the reaction between two very reactive elements to make the everyday table salt you can safely sprinkle on foodstuffs.

Sodium chloride crystals. Sodium chloride is soluble in water. The chemical industry uses an electric current to convert sodium chloride solution into chlorine, hydrogen, and sodium hydroxide.

A chemical reaction in pictures: sodium and chlorine react to make sodium chloride.

Salts

Salts such as sodium chloride are crystalline. The crystals of sodium chloride are shaped like cubes. So are the crystals of calcium fluoride shown on page 44 at top left.

Salts have much higher melting and boiling points than compounds such as chlorine and bromine, which are made up of small molecules.

Crystals of the mineral galena, which is an ore of lead. Galena consists of insoluble lead sulfide.

Chemical	Formula	Melting point (°C)	Boiling point (°C)
sodium	Na	98	890
chlorine	Cl_2	−101	−34
sodium chloride	NaCl	808	1465
potassium	K	63	766
bromine	Br_2	−7	58
potassium bromide	KBr	730	1435

Sodium chloride is an example of a salt that is soluble in water. There are many other examples of soluble salts, including most of the compounds of alkali metals with halogens.

Some salts are insoluble in water. Lithium fluoride is an example of a salt which is only very slightly soluble in water. Many minerals consist of insoluble salts. Fluorite (CaF_2) is one example. Others are galena (PbS) and the brassy looking pyrites (FeS_2), sometimes called fool's gold.

Crystals of the mineral pyrites. Pyrites consist of insoluble iron sulfide.

Molten salts and electricity

The apparatus on the right is used to investigate whether or not chemicals conduct electricity. The crucible contains some white powdered solid. This is lead bromide.

At first the bulb does not light, showing that the solid does not conduct electricity. This is true of all compounds of metals with non-metals; they do not conduct when solid.

Heating the crucible melts the lead bromide. As soon as the compound is **molten**, there is a reading on the meter. This shows that a current is flowing round the circuit. As a liquid, the compound is a conductor. That is not all. The electric current causes the compound to decompose chemically. The most obvious change is the bubbling around the positive **electrode**. Puffs of orange gas appear as the bubbles burst. The orange gas is bromine.

After a while, it is possible to show that lead has formed at the negative electrode. This is done by switching off the current and pouring the liquid from the crucible into a mortar. The unchanged lead bromide quickly solidifies. Gently crushing the solid reveals a shiny lump of metallic lead. So the electric current splits the compound into its elements: lead and bromine.

Salts in solution and electricity

Soluble salts also conduct electricity. This can be studied using the apparatus shown on the right. There are changes at the electrodes when an electric current flows.

The presence of water has an effect on the chemicals produced when a salt solution conducts electricity. The products are not always the same as the elements in the compound.

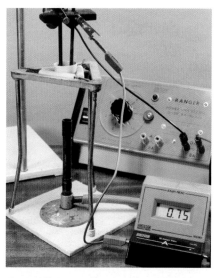

The crucible contains lead bromide. The carbon rods dipping into the crucible are the electrodes. A current begins to flow in the circuit when the lead bromide is hot enough to melt.

This apparatus is used to study the changes at the electrodes when a solution of a salt conducts electricity. In the example shown, the flow of an electric current is producing gases at the electrodes.

Questions

1 Draw up a table to compare the properties of sodium, chlorine, and sodium chloride.

2 Refer to the table of data on page 54. Which of the chemicals is a liquid:
  a at room temperature?
  b at the boiling point of water?
  c at 1000 °C?

3 Draw a two-dimensional line diagram and circuit diagram to represent the apparatus used to show that lead bromide conducts electricity when hot enough to melt.

Key words
molten
electrode

Find out about:
▶ ions
▶ ionic compounds
▶ explaining properties of salts

Michael Faraday lectured at the Royal Institution. He started the Christmas lectures, which continue today in the same lecture theatre.

Key words
electrolysis
ions
positive ions
negative ions

J   Ionic theory

Electrolysis

An electric current can split a salt into its elements if it is molten or dissolved in water. This is the process called **electrolysis**. The term 'electro-lysis' is based on two Greek words that mean 'electricity-splitting'.

The discovery of electrolysis was very important in the history of chemistry because it made it possible to split up compounds which previously no-one could decompose. An English chemist, Humphry Davy, was Professor of Chemistry at the Royal Institution in London from 1802 to 1812. During 1807 and 1808 he used electrolysis to isolate for the first time the elements potassium, sodium, barium, strontium, calcium, and magnesium.

Faraday's theory

Michael Faraday also worked at the Royal Institution. He began as an assistant to Humphry Davy but established himself as a leading scientist in his own right. In 1833 he began to study the effects of electricity on chemicals.

Faraday decided that compounds that can be decomposed by electrolysis must contain electrically charged particles. Since opposite electrical charges attract each other, he could imagine the negative electrode attracting positively charged particles and the positive electrode attracting negatively charged particles.

The charged particles move towards the electrodes. When they reach the electrodes, they turn back into atoms. This accounts for the chemical changes that decompose a compound during electrolysis.

Faraday consulted a Greek scholar, and together they named the moving, charged particles **ions**, from a Greek word meaning 'wanderer'.

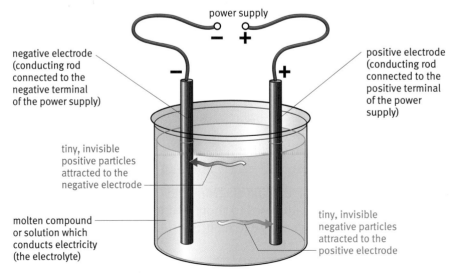

A modern outline of Faraday's ionic theory

H Explaining electrolysis

Chemists continue to use ionic theory to explain electrolysis. According to the theory, salts such as sodium chloride consist of ions.

Sodium chloride is made up of sodium ions and chloride ions. Sodium ions, Na^+, are positively charged. The chloride ions, Cl^-, carry a negative charge. These oppositely charged ions attract each other.

A crystal of sodium chloride consists of millions and millions of Na^+ and Cl^- ions closely packed together. In the solid, these ions cannot move towards the electrodes, and so the compound cannot conduct electricity. The ions can move when sodium chloride is hot enough to melt or when it is dissolved in water.

During electrolysis the negative electrode attracts the **positive ions**. The positive electrode attracts the **negative ions**. When the ions reach the electrodes, they lose their charges and turn back into atoms.

Metals form positive ions, and non-metals generally form negative ions.

Elements and compounds

Ionic theory can help to explain why compounds are so different from their elements. Sodium atoms are dangerously reactive, and so are chlorine molecules. Sodium chloride is safe because its ions are much less reactive.

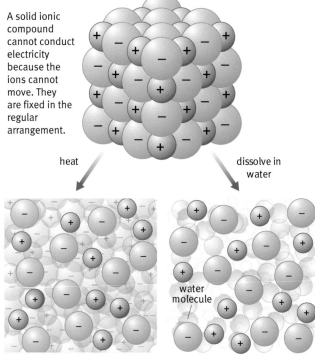

A solid ionic compound cannot conduct electricity because the ions cannot move. They are fixed in the regular arrangement.

heat

dissolve in water

When an ionic compound is heated strongly, the ions move so much that they can no longer stay in the regular arrangement. The solid melts. Because the ions can now move around independently, the molten compound conducts electricity.

When an ionic compound has dissolved, it can conduct electricity because its ions can move independently among the water molecules.

water molecule

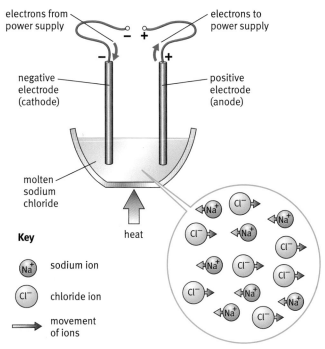

electrons from power supply

electrons to power supply

negative electrode (cathode)

positive electrode (anode)

molten sodium chloride

heat

Key

Na^+ sodium ion

Cl^- chloride ion

⟶   movement of ions

Sodium chloride conducts when molten because its ions can move towards the electrodes.

Questions

1 Why do solid compounds made of ions not conduct electricity?

2 Chemists sometimes call the negative electrode the cathode. Cations are the ions that move towards the cathode. What is the charge on a cation? Which type of element forms cations? Give an example of a cation.

3 Chemists sometimes call the positive electrode the anode. Anions are the ions that move towards the anode. What is the charge on an anion? Which type of element forms anions? Give an example of an anion.

Find out about:
- atoms and ions
- electron configurations of ions
- the formulae of ionic compounds

к Ionic theory and atomic structure

Atoms into ions

Faraday could not explain how atoms turn into ions because he was working long before anyone knew anything about the details of atomic structure. Today, chemists can use the shell model for electrons in atoms to show how atoms become electrically charged.

The metals on the left-hand side of the periodic table form ions by losing the few electrons in the outer shell. This leaves more protons than electrons, and so the ions are positively charged.

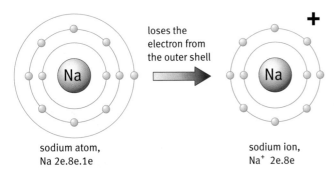

loses the electron from the outer shell

sodium atom,
Na 2e.8e.1e

sodium ion,
Na$^+$ 2e.8e

A sodium atom turns into a positive ion when it loses a negatively charged electron.

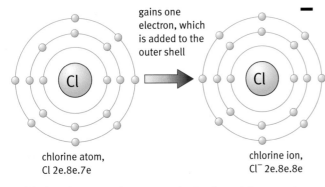

gains one electron, which is added to the outer shell

chlorine atom,
Cl 2e.8e.7e

chlorine ion,
Cl$^-$ 2e.8e.8e

A chlorine atom turns into a negative ion by gaining an extra negatively charged electron.

All the metals in group 1 have one electron in the outer shell. The diagram on page 51 shows that removing the first electron from a sodium atom needs relatively little energy. The same is true for the other group 1 metals, so they all form ions with a 1+ charge: Li$^+$, Na$^+$, and K$^+$, for example.

Chlorine gas consists of Cl$_2$ molecules. But it is easier to see what happens when chlorine gas forms by looking at one atom at a time, as shown in the diagram on the left. As each chlorine atom turns into an ion, it gains one electron and becomes negatively charged, Cl$^-$.

Electron configurations of ions

Notice that when sodium and chlorine atoms turn into ions, they end up with the same electron configuration as the nearest noble gas in the periodic table. This is generally true for simple ions of the first 20 or so elements in the periodic table.

An explanation of why this is so requires a detailed analysis of all the energy changes when metals react with non-metals. This is something you will study if you go on to a more advanced chemistry course.

Ions into atoms

Electrolysis turns ions back into atoms. Metal ions are positively charged, so they are attracted to the negative electrode. It is a flow of electrons from the battery into this electrode that makes it negative. Positive metal ions gain electrons from the negative electrode and turn back into atoms

Non-metal ions are negatively charged, so they are attracted to the positive electrode. This electrode is positive because electrons flow out of it to the battery. Negative ions give up electrons to the positive electrode and turn back into atoms.

H Formulae of ionic compounds

The formula of sodium chloride is NaCl because there is one sodium ion (Na^+) for every chloride ion (Cl^-). There are no molecules in everyday table salt, only ions.

Not all ions have single positive or negative charges like sodium and chlorine. The formula of lead bromide is $PbBr_2$. In this compound there are two bromide ions for every lead ion. All compounds are overall electrically neutral, so the charge on a lead ion must be twice that on a bromide ion. A bromide ion, like a chloride ion, has a single negative charge (Br^-), so a lead ion must have a double positive charge (Pb^{2+}).

Compound	Ions present		Formula
	Positive	Negative	
magnesium oxide	Mg^{2+}	O^{2-}	MgO
calcium chloride	Ca^{2+}	Cl^-	$CaCl_2$
		Cl^-	
aluminium oxide	Al^{3+}	O^{2-}	Al_2O_3
	Al^{3+}	O^{2-}	
		O^{2-}	

Examples of formulae of ionic compounds

Ions in the periodic table

The charges of simple ions show a periodic pattern. You can see this from the diagram below, in which the ionic symbols appear in the periodic table. Many of the transition metals in the middle block of the table can form more than one type of ion. Iron, for example, can form Fe^{2+} and Fe^{3+} ions, while copper can exist as Cu^+ and Cu^{2+} ions. Why this should be so is something you will study if you go on to a more advanced chemistry course.

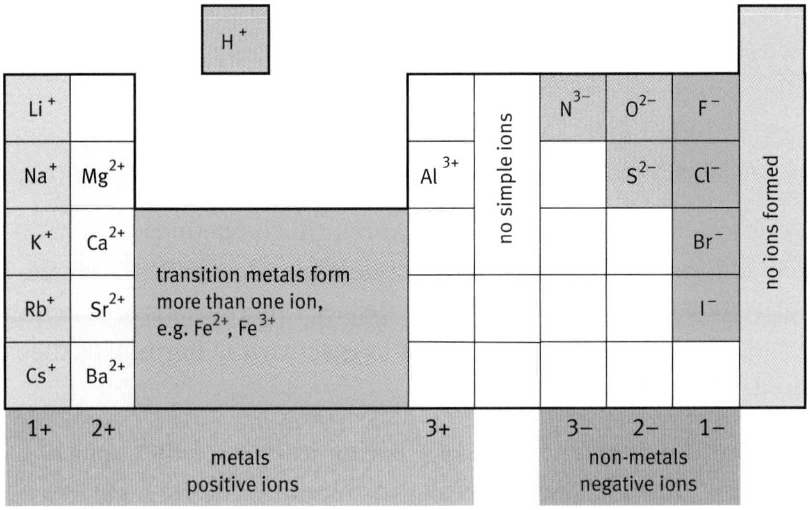

Simple ions in the periodic table

Questions

1 Draw diagrams to show the number and arrangement of electrons in a lithium atom and in a lithium ion. What is the charge on a lithium ion?

2 Draw diagrams to show the number and arrangement of electrons in a fluorine atom and in a fluoride ion. What is the charge on a fluoride ion?

3 Write down the electron configurations of:
   a a fluoride ion (nucleus with 9 protons and 10 neutrons)
   b a neon atom (nucleus with 10 protons and 10 neutrons)
   c a sodium ion (nucleus with 11 protons and 12 neutrons). In what ways are a fluoride ion, a neon atom, and a sodium ion the same. How do they differ?

4 With the help of the table of ions, work out the formulae of these ionic compounds:
   a potassium iodide
   b calcium bromide
   c aluminium chloride
   d magnesium nitride
   e aluminium sulfide

Find out about:
▶ atoms, molecules, and ions
▶ chemical species

L Chemical species

In this module you have met the idea that the same element can take different chemical forms with distinct properties. Chemists describe these different forms as **chemical species**.

Species of chlorine

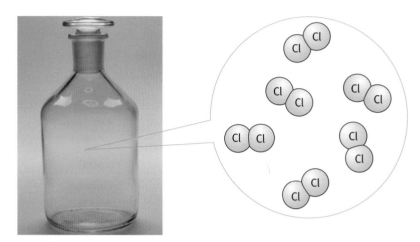

Chlorine gas consists of chlorine molecules. Chlorine molecules are chemically very reactive.

Chlorine has three simple species: atom, molecule, and ion. Each of these species of chlorine has distinct properties. Chlorine atoms (Cl) do not normally exist in a free state. They rapidly pair up to form chlorine molecules (Cl_2). However, ultraviolet radiation can split chlorine and chlorine compounds into atoms. This is what happens to CFCs such as CCl_3F when they get into the upper atmosphere. In the full glare of the Sun's radiation the molecules break up into atoms. Then the very reactive free chlorine atoms rapidly destroy ozone. Lowering the concentration of ozone creates the so-called 'hole' in the ozone layer.

Chlorine gas at room temperature consists of chlorine molecules (Cl_2). These are very reactive, as illustrated by the chemistry of the alkali metals and halogens described in section D. The chlorine molecules are reactive enough to do damage to human tissues, so the gas is given the label 'toxic'.

Chloride ions are quite different. They occur in compounds such as sodium chloride and magnesium chloride. Chloride ions in these salts are essential to life and occur in all living tissues. Chloride ions are chemically active in many ways, but they are not as reactive and harmful as the atoms or molecules of the element.

There are more complex species of chlorine with the element joined to other atoms. This includes molecules that contain chlorine and other elements, such as tetrachloromethane (CCl_4).

Key words
chemical species

Species of sodium

There are only two species of sodium: atom and ion. The atoms in sodium metal are chemically very active. In the presence of other chemicals, the sodium atoms react to produce compounds containing sodium ions.

Sodium combines with chlorine to produce the ionic compound sodium chloride. This is made up of two chemical species: Na^+ and Cl^-. These two ions are quite unreactive. Sodium chloride is soluble in water, but its solution is neutral. Water does not react with the ions.

When sodium reacts with water, it produces another ionic compound: sodium hydroxide, Na^+ and OH^-. The sodium hydroxide dissolves in the water to give a solution that is very alkaline. It is the hydroxide ions that make a solution of sodium hydroxide alkaline, not the sodium ions. You will find out more about the ionic theory of acids and alkalis in Module C6 *Chemical synthesis*.

Chemical species of the natural environment

You will learn more about the importance of being precise about chemical species in Module C5 *Chemicals of the natural environment* when you explore the chemical changes that affect elements such as nitrogen in the environment. Nitrogen gas is very different from oxides of nitrogen, which in turn are quite different in their properties from a salt containing nitrate ions.

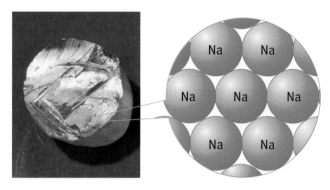

Sodium metal consists of sodium atoms. Sodium atoms are chemically very reactive.

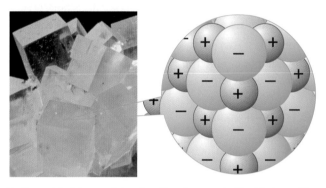

Sodium chloride consists of sodium ions and chloride ions. These ions are not very reactive.

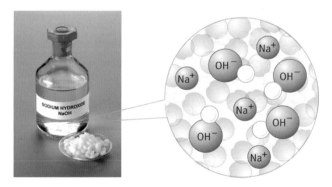

Sodium hydroxide is a strong alkali. It consists of sodium ions and hydroxide ions. In solution the ions move around separately mixed with water molecules.

Questions

1 Use the idea of chemical species to explain why the properties of sodium chloride are very different from the properties of its elements sodium and chlorine.

2 Give the name and formulae of all chemical species in:
   a potassium
   b bromine
   c potassium bromide

3 a Identify four distinct chemical species in unpolluted air by giving their names and formulae.
   b Identify three more chemical species present in the polluted air of a busy city street.

C4 Chemical patterns

Summary

You have now met some of the key patterns and theories that chemists use to make sense of the world and to explain how roughly 100 elements can give rise to such a huge variety of chemical compounds.

Atomic structure and the periodic table

- Atoms have a tiny central nucleus surrounded by negative electrons.

- The chemistry of an element is largely determined by the number and arrangement of the electrons in its atoms.

- The number of electrons is equal to the proton number of the atom.

Electrons in atoms

- Electrons in an atom have definite energies.

- The electron shell with the lowest energy fills first until it contains as many electrons as possible, then the next shell starts to fill.

Periodic table

- Arranging the elements in order of their proton numbers gives rise to the periodic table.

- For the first two rows in the table each period corresponds to the filling of an electron shell.

- After calcium the relationship between atomic structure and the periodic table becomes more complex and gives rise to the block of transition elements.

Groups

- Each column in the periodic table consists of a group of related elements.

- The elements in a group have similar chemistries because they have the same number of electrons in the outer shell.

- There are trends in the properties of the elements down a group because of the increasing number of inner full shells.

Atoms into ions

- When metals react with non-metals, the metal atoms lose electrons while the non-metal atoms gain electrons.

- This produces ionic compounds such as sodium chloride, Na^+Cl^-.

- The properties of an ionic compound are the properties of its ions, which behave in a different way from the atoms or molecules in the elements.

Questions

1 Copy this table and extend it to include the first 20 elements.

Element	Proton number	Number of electrons in each shell			
		First shell	Second shell	Third shell	Fourth shell
hydrogen	1	1			
helium	2	2			
lithium	3	2	1		

2 Summarize in outline how the model of atomic structure with electrons in shells can account for:

a the line spectra of elements

b the arrangement of the elements in the periodic table

c the charges on simple ions.

3 The table below shows part of the periodic table with only a few symbols included.

group / period	1	2		3	4	5	6	7	8
1									He
2								O	
3	Na	Mg		Al	Si		S	Cl	
4						Ge		Br	
5									
6	Cs							As	

a Using only the elements in the table, write down the symbols for the following:

i a metal stored in oil

ii two non-metals that are gases at room temperature

iii an element used to kill bacteria

iv a metal that floats on water

v an element with similar properties to silicon (Si)

vi an element that has molecules made up of two atoms

vii the element with the largest number of protons in the nucleus of its atoms

viii the most reactive metal

ix an element with two electrons in the outer shell of its atoms

x an element X that forms an ionic chloride with the formula XCl

b Predict the formula of the compound of sulfur (S) with hydrogen.

c Is astatine (As) a solid, liquid, or gas at room temperature? Explain how you decide on your answer.

4 Potassium chloride (KCl) melts at 772 °C and boils at 1407 °C. Crystals of potassium chloride are colourless and shaped like cubes. The compound is soluble in water. A solution of potassium chloride conducts electricity.

a Is potassium chloride a solid, liquid, or gas at room temperature?

b Why do crystals of KCl all have the same shape?

c Why does a solution of potassium chloride conduct electricity?

d State two ways that potassium chloride differs from the element potassium.

e State two ways that potassium chloride differs from the element chlorine.

f Why are the properties of potassium chloride so different from its elements?

Why study motion?

Humans have always been interested in how things move and why they move the way they do. Motion is such an obvious part of our everyday lives that we cannot really claim to know very much about the natural world if we cannot explain and predict how objects move.

The science

One tantalizing thought has always driven people who have studied motion – is it possible that every example of motion we observe can be explained by a few simple rules (or laws) that apply to everything? Remarkably the answer is 'yes'. And these laws are so exact and precise that they can be used to predict the motion of an object very precisely. A key idea is force. A force acting on an object, unless it is cancelled out by another force, causes a change in its motion.

Physics in action

Understanding forces and motion has enabled scientists and engineers to design and build more efficient cars and trains, to develop aircraft, and to fly spacecraft with enormous accuracy to the furthest reaches of the Solar System. The same science ideas are used in testing new materials. Although our understanding of motion was developed by Isaac Newton in the 17th century, it is still at the heart of many innovations and developments in the 21st century.

Explaining motion

Find out about:

- how forces always arise from an interaction between two objects
- friction and reaction of surfaces
- instantaneous and average speed, and velocity
- the idea of momentum, and how the momentum of an object changes when a force acts on it
- everyday examples of motion, including the principles on which traffic safety measures are based
- gravitational potential energy and kinetic energy

Find out about:

▶ how forces arise when two objects interact

▶ contact and action-at-a-distance forces

A Forces in all directions

To start anything moving requires a **force**. A firework explodes because of a chemical reaction inside. The forces on the sparks in the starburst are a result of the chemical reaction. The firework fragments in the photograph on the left are being pushed out in all directions. The force on each fragment can be represented by an arrow, as shown below the photograph. In this case, there are lots of fragments and lots of forces.

How do forces arise?

To help understand how forces arise, it is easier to start with a simpler situation where there are only two objects moving apart.

Sophie and Sam have gone out sailing on the lake, but the wind has dropped. Their boats are together but neither is moving. They have no oars. They do not know what to do to get back to land.

The chemical reaction inside the firework shell produces forces that send the burning fragments out in all directions, producing a sphere of shooting stars.

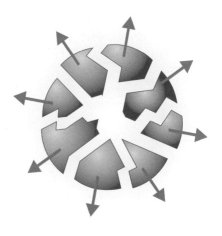

The size of the force on each fragment is shown by the length of the arrow.

Two stationary boats and no oars: how can Sophie or Sam get moving?

Perhaps one of them could push the other. But when they do this, they both move. However hard they try, Sam and Sophie are unable to make only one of the boats move. It is not possible for them to push so that one experiences a force but not the other.

You can tell something very important about forces from this:

▶ Forces always arise from an **interaction** between two objects.

So forces always come in pairs. The two forces in an **interaction pair** are

▶ equal in size

▶ opposite in direction

This is always true. And it does not depend on the size or strength of the two people involved. Another important thing to notice is that

▶ the two forces act on different objects

In this example, one force of the pair acts on Sam and the other on Sophie.

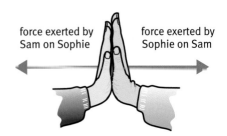

force exerted by Sam on Sophie   force exerted by Sophie on Sam

Forces always arise in pairs. Here Sophie pushes Sam and experiences a force in return.

'Things' can push (and pull) too!

You are used to the idea that people and animals can push and pull. You also know that machines and motors can exert forces. But in fact anything can exert a force if it is involved in an interaction.

The diagram on the right shows Deborah, a roller-skater. She pushes against the wall and immediately starts to move backwards. When she pushes the wall, the wall pushes back on her. It pushes back with an equal force in the opposite direction. This force makes Deborah start to move.

Action at a distance

Where the interacting objects touch each other, the forces are known as contact forces. There are also forces that act at a distance. The forces caused by gravity and magnetism are examples. However, *all* forces arise from interactions.

An apple falls from a tree because of the force pulling it downwards, towards the centre of the Earth. But gravity is an attraction between two objects. So the apple also exerts an equal and opposite force on the Earth! This does not have much effect, however, because the Earth is so large.

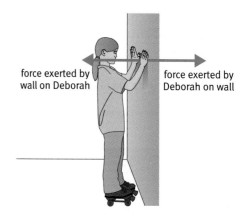

force exerted by wall on Deborah

force exerted by Deborah on wall

Deborah pushes against the wall. The other force in the interaction pair then starts her moving on her roller-skates.

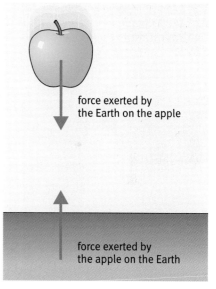

force exerted by the Earth on the apple

force exerted by the apple on the Earth

A gravitational interaction – again forces always arise in pairs.

Questions

1 List four examples of interaction pairs of forces mentioned on these pages.

2 What three things are always true about interaction pairs?

③ These two ring magnets are repelling each other. Notice that both magnets are being pushed aside.

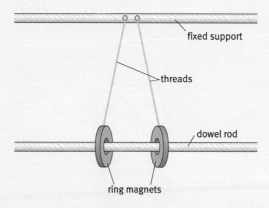

fixed support

threads

dowel rod

ring magnets

How could you modify this apparatus to show that *attraction* forces between magnets also arise in pairs? Sketch how you would set it up, and write down what you would expect to see.

Key words

force
interaction
interaction pair

Find out about:

▶ the forces that enable people and vehicles to get moving
▶ rockets and jet engines

B How things start moving

Rockets

When objects explode, the pieces usually travel outwards in all directions. But if an object is designed so that it does not break up, and everything that comes out of it goes in one direction, then you have a rocket.

The photograph on the left shows one of the most famous rocket launches: the *Apollo 11* mission to land the first people on the Moon. The interaction pair of forces is shown on the photo. Burning hot gases are pushed out of the base of the rocket, and the rocket is pushed in the opposite direction.

Rockets carry with them everything they need to make the burning gases they push against. This means that they can work in space as well as in air.

Jet engines

Jet engines use the same basic idea as rockets. Air is drawn into the engine and pushed out at the back. The other force of the pair pushes the engine forward. Jet engines need to draw air in, so they cannot work in space.

How does a car get moving?

To make a car move, the engine has to make the wheels turn. This causes a forward force on the car. To understand how, think first about a car trying to start on ice. If the ice is very slippery, the wheel will just spin. The car will not move at all. The spinning wheel produces no forward force on the car. Now imagine a car on a muddy track. The rally car below is throwing up a shower of mud as it tries to get going.

force exerted on the rocket

force exerted on the exhaust gas

The start of the longest journey humans have made so far – to the Moon. A huge force is needed to push a rocket like this upwards. It is provided by the hot exhaust gases, which are formed by burning the fuel.

As it rotates, the wheel exerts a force backwards on the ground – with dramatic results in this case!

You can see that there is an interaction between the wheel and the ground. The wheel is causing a backwards force on the ground surface. This makes the mud fly backwards. Mud, however, moves when the force is quite small. The other force of the interaction pair is the forward force on the car. It is equal in size. So it is also small – and not big enough to get the car moving.

Now imagine a good surface and good tyres, which do not slip. Again, the engine makes the wheel turn. It pushes back on the road. The wheel cannot slip, and it exerts a very large force backwards on the road surface. So the other force of the interaction pair is the same size. It is this large forward force which gets the car moving.

Questions

1 Jet engines are suitable for aircraft but not for travel in space. Explain why. How do rockets overcome this problem?

2 A boat propeller pushes water backwards when it spins round. Use the ideas on these pages to write a short paragraph explaining how this makes the boat move forward. Draw a diagram and label the main forces involved, to illustrate your explanation.

3 Sketch a matchstick figure walking. Mark and label the interaction pair of forces on the foot in contact with the ground.

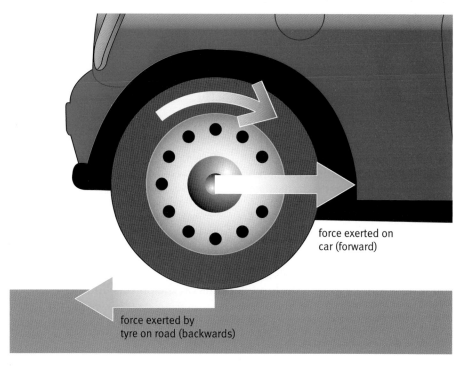

force exerted on car (forward)

force exerted by tyre on road (backwards)

If the tyre grips the road and does not spin, the second force of the interaction pair results in a forward force on the axle. This pushes the car forward.

Walking

When you walk, you push back on the ground with each foot in turn. The ground then pushes you forward. You are not usually aware of this. If you tiptoe carefully across a floor, it does not feel as though you are pushing backwards on it. You only become aware of the importance of this interaction when the surface is slippery – when you try to walk on an icy surface, for example. Because you cannot push it back, it is unable to move you forward.

Icy surfaces are difficult to walk on. Your foot cannot get a grip to push back on the surface, so the surface does not push you forwards.

Find out about:

▶ friction and what causes it

c Friction – a responsive force

Friction often seems a nuisance. But without it, you could not start walking. Cars could not get moving.

What is friction?

Jeff is a workman. He is trying to push a large box along a level floor. Think about the forces involved:

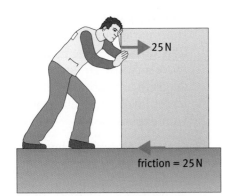

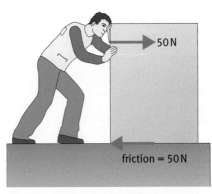

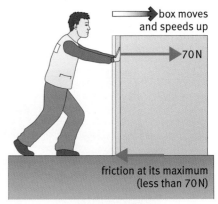

1. Jeff pushes the box with a force of 25 N, to try to slide it along. It does not move. The friction force exerted by the floor on the box is 25 N. This exactly balances Jeff's push.

2. Jeff then pushes harder, with a force of 50 N. The box still does not move. The friction force exerted by the floor on the box is now 50 N. Again, this balances Jeff's push.

3. Jeff pushes harder still, exerting a force of 70 N. The box starts to move. It keeps speeding up while Jeff pushes. 70 N is bigger than the maximum possible friction force for this box and floor surface.

So friction is an unusual force. It adjusts its size in response to the situation – up to a limit. This limit depends on the objects and surfaces involved.

What causes friction?

Friction is a common type of force. But surprisingly, scientists do not yet agree on an explanation of the friction force between two sliding surfaces. Some things about friction are, of course, understood. It has to do with the roughness of the surfaces. Even surfaces that seem smooth have quite large humps and hollows if you look at them under a microscope.

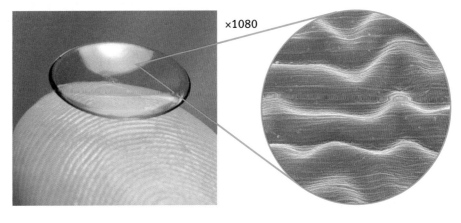

Even the smoothest surface is really quite rough. At the microscopic level, it has humps and hollows. The photograph on the right shows the surface of a contact lens magnified 1080 times.

When two surfaces are put together, the bumps on one can fit into the hollows of the other. When one object slides over another, it has to ride up and down over these bumps. To see why this requires a force, think about trying to slide two brushes past each other. The bristles on each brush will exert a sideways force on the other one.

But there is more to it than this. Because all surfaces are really quite rough, they only touch at a few points. They touch where a bump on one meets a bump on the other. So there are only a few real points of contact. As a result, the pressure at these points is very large. It is large enough to 'cold weld' them together. So when you slide one object over another, you have to keep breaking these tiny welds. And this needs a force.

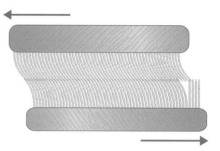

You can see how the friction force arises when you try to slide two brushes past each other.

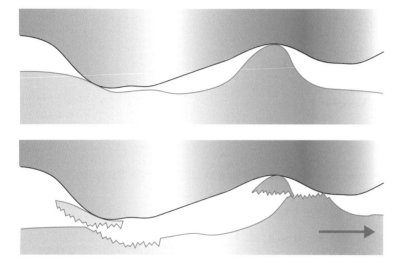

The force of friction arises because lots of tiny welds have to be broken while the objects slide past each other. With modern methods of detection, it is possible to show that tiny bits of one surface stick to the other, after they have been slid across each other.

Key words
friction

Questions

1 List three everyday situations in which we try to reduce friction, and three where we try to make friction as large as possible.

2 Use the ideas on these pages to write short explanations of the following observations:

a We can reduce the friction between two surfaces by putting oil on them.

b It is easier to push a box across the floor when it is empty than when it is full.

c Sometimes, when you polish two surfaces, the friction force between them gets bigger.

3 Scientists think that tiny 'spot welds' occur when two surfaces are in contact. What is the evidence for this?

4 Sketch three diagrams of the workman, Jeff, pushing the box on page 70. Mark and label the forces acting *on Jeff* as he pushes. Use the length of each force arrow to indicate the size of the force.

Find out about:

▶ how a surface exerts a reaction force on any object that presses on it

D Reaction of surfaces

If you hold a tennis ball at arm's length and let it go, it immediately starts to move downwards. There is a force acting on the tennis ball. This force is the pull exerted on it by the Earth. It is due to the interaction known as gravity.

But if you put a tennis ball on a table so that it does not roll about, it does not fall. The force of gravity has not suddenly stopped or been switched off. There must be another force that cancels it out. The only thing that can be causing this is the table. The table must exert an upward force on the ball that balances the downward force of gravity.

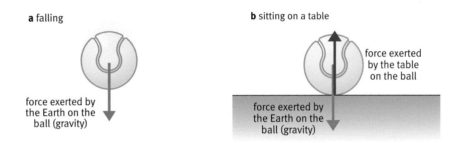

a falling

force exerted by the Earth on the ball (gravity)

b sitting on a table

force exerted by the table on the ball

force exerted by the Earth on the ball (gravity)

The forces acting on a tennis ball **a** falling and **b** sitting on a table.

How can a table exert a force?

Although it may seem strange, tables can and do exert forces. To understand how, imagine an object, like a school bag, sitting on the foam cushion of a sofa. The bag presses down on the foam, squashing it a bit. Because foam is springy, it then pushes upwards on the bag, just like a spring. Like a spring, the more it is squeezed, the harder it pushes back. So the bag sinks into the foam until it reaches the point where the push of the foam on it exactly balances the downward pull of gravity on it.

The same thing happens, though on a much smaller scale, when the bag sits on a table top. A table top is not so easily squeezed as a foam cushion. But it *can* be squashed. This may not be visible to the naked eye, however. We call this upward force which a hard surface exerts when something presses on it the **reaction** of the surface.

Of course, a table cannot always exert an upward force on an object to balance the downward force on it. There is a limit. This limit depends on the material the table is made from. If the force exerted on the table top gets bigger than this, it is distorted beyond the point where it can spring back. It then breaks. Up to this point, however, it exerts an upward force that exactly matches the downward force exerted on it.

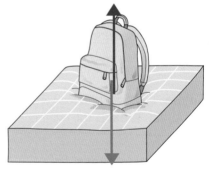

force exerted by the cushion on the bag (reaction)

force exerted by the Earth on the bag (gravity)

The bag squeezes the foam until the upward force of the springy foam on the bag exactly balances the downward gravity force on the bag.

E Adding forces

Find out about:
▶ how to add the forces acting on an object

The discussion on pages 70-72 about friction and reaction of surfaces used an idea that may seem obvious:

▶ If there is a force acting on an object, but it is not moving, then there must be another force balancing (or cancelling out) the first one.

For Jeff, the workman pushing the box (page 70), the other force is friction. For a bag sitting on a table, the other force is the reaction of the table surface.

If the forces acting on an object balance each other, we say they add to zero. Adding several forces that act on the same object is straightforward. But you must take the direction of each force into account. The sum of all the forces acting on an object is called the **resultant force**. The diagrams below show some examples.

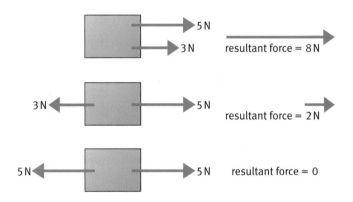

To find the resultant force acting on an object, you add the separate forces. You must take account of their directions.

Key words
reaction (of a surface)
resultant force

Questions

1 What happens to a 'hard' surface when something sits on it? Is it really as hard as it seems?

2 Look back at the diagram on page 67 of Deborah pushing against a wall. She moves because the wall exerts a force on her. Use the ideas on these pages to write a paragraph explaining what happens to the wall at the point where Deborah pushes – and how the wall is therefore able to exert a force on her.

3 Imagine the bag in the diagram on page 72 hanging from a string. The string must be exerting an upward force on the bag, equal to the downward force of gravity on it. How does the string exert this upwards force? Use the ideas on these pages to suggest an explanation.

Find out about:
- ▶ how to calculate the speed of a moving object
- ▶ catching speeding motorists

F How fast are you going?

Cars have speedometers. But how can you tell how fast you are going on your bike?

Most bicycles do not have speedometers, so you cannot measure your speed directly. But you can time how long it takes to cycle between two places – two lamp-posts, for example. And you can measure how far you have cycled – the distance between the two lamp-posts. Then you can work out the **average speed**. Here's how.

Key words
average speed
instantaneous speed

Average speed

To calculate average speed is quite easy. You use the equation

$$\text{average speed} = \frac{\text{distance travelled}}{\text{time taken}}$$

However, knowing the average speed is sometimes not very useful. For most journeys, the speed is not always the same. It varies. For instance, imagine you are going to drive to a friend's house:

Your journey

1 Town traffic

During the first part of your journey, you drive from home to the motorway. This takes you through busy city streets. The speed limit is 30 mph (miles per hour), but often you are travelling slower than this.

2 Motorway

You travel on the motorway for 1 hour. In that time, you go 60 miles. So your average speed on the motorway is

$$\frac{60 \text{ miles}}{1 \text{ hour}} = 60 \text{ mph}$$

3 Country lane

You turn off the motorway. You have 6 more miles to go. But you get held up on a narrow country road. It takes you 30 minutes (0.5 hour)! So your average speed on this part of the journey is

$$\frac{6 \text{ miles}}{0.5 \text{ hour}} = 12 \text{ mph}$$

4 End of journey

You look at your watch and the mileometer when you arrive. The whole journey of 76 miles has taken you exactly 2 hours. So, for the whole journey, your average speed was

$$\frac{76 \text{ miles}}{2 \text{ hours}} = 38 \text{ mph}$$

So, you may know the average speed for the whole journey. But you cannot tell anything about how fast the car was going at any particular moment. If you did drive steadily at 38 mph (the average speed) for the whole journey, it would take you exactly 2 hours. But in practice, you did not. Your speed kept changing.

Instantaneous speed

The speed at a particular moment is called the **instantaneous speed**. If you were able to calculate average speeds over shorter and shorter time intervals, these would get closer and closer to the instantaneous speed of the car. In practice, to estimate the instantanous speed, you measure the average speed over a very short time interval.

The speedometer in a car shows the driver the instantaneous speed.

Questions

1 a Your whole journey above was 76 miles. How far was it from home to the motorway?

  b Altogether, you drove for 2 hours. How long did it take you to reach the motorway?

  c So, what was your average speed in miles per hour from home to the motorway?

Catching the speeders

The police have several methods they can use to measure a vehicle's speed:

① **Gatso speed cameras** use radar (see method 3 below) to detect vehicles that are above the speed limit. The camera then takes two photographs of the vehicle, half a second (0.5 s) apart, to provide evidence. Distance markers on the road (they are 1.5 metres apart here) show how far the car has travelled in this time.

detector cables

② **Truvelo speed cameras** are triggered by detector cables in the road. Pressure sensors in the cables detect when a car is passing over. A computer in the camera measures the speed of passing cars by recording the time the car takes to travel from one cable to the next. If it is going faster than the speed limit, a picture is taken. These cables are 10 cm (0.1 m) apart.

③ **Police radar guns** bounce microwaves off approaching cars. Microwaves reflected off an oncoming car have a higher frequency than the original waves. These are picked up again by the radar gun. The gun uses the change in frequency to calculate the instantaneous speed of the car.

Questions

2 Look at the car moving away from you in the top two photographs.

  a Estimate the distance that it moves between the top two photographs.

  b Estimate the speed of this car in metres per second.

  c Is this its average speed or its instantaneous speed? Explain your answer.

3 The detector cables in the bottom left photograph record a time of 0.008 s for this white van. The speed limit is 13 metres per second (30 mph).

  a Is the van above the speed limit?

  b Does this method measure the van's average speed or its instantaneous speed? Explain your answer.

④ You may have seen signs that display a car's speed as it enters a 30 mph zone – with a warning to 'Slow down!'. Which of these methods of measuring speed do you think these 'shame displays' are most likely to use? Explain why.

G Picturing motion

You can use a graph to describe a journey more easily than with words:

▶ a distance–time graph shows how far a moving object is from its starting point at every instant during its journey

▶ a speed–time graph shows the speed of the moving object at every instant during its journey

Graphs do not only summarize information about the motion. They also help analyse it. This is how.

Distance–time graphs

On the left is a distance–time graph for the motorway section of the car journey described on page 74. Try taking readings from this graph of the distance the car has travelled after a quarter of an hour, half an hour, and one hour. The distance increases steadily as time goes by. The constant slope of the distance–time graph indicates a steady speed.

The distance–time graph below shows a cycle ride that Vijay did during his school holidays. The graph has four sections. In each section, the slope of the graph is constant. This means that his speed is constant during that section of the ride.

▶ In the first hour, Vijay travels 15 miles. His speed is 15 mph.

▶ In the second hour, he only travels 5 miles because of the strong wind. The shallower slope indicates a lower steady speed.

▶ In the third section (from 2.0 to 2.5 hours), his distance from home does not change. He is stopped. This is what a horizontal section of a distance–time graph means.

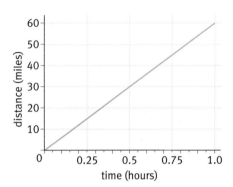

Distance–time graph for a car journey along the motorway

Questions

1 a By looking at the slope of the graph on the right, say how Vijay's speed in the final section of his ride compares with his speed in the first and second sections.

b How far does Vijay travel during the final section of his journey (from 2.5 to 3.5 hours)? So what is his speed during this section?

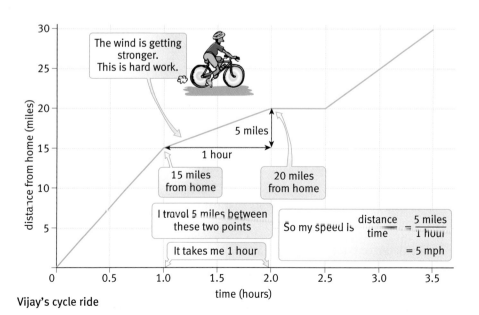

Vijay's cycle ride

Real journeys, however, do not consist of sections at steady speed. Instead the speed is always changing. And these changes may be gradual. More realistic graphs have slopes that change smoothly.

Look at the graph on the right. It shows a car journey. The slope tells you the speed of the car. So if the graph goes up ever more steeply, then the car must be speeding up, or accelerating. And if the slope is decreasing, then so is the speed. The final part of this distance–time graph slopes downwards. This shows that the distance from the starting point is getting less. The car is moving in the opposite direction, back towards the start.

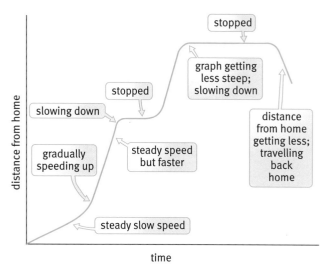

A more realistic distance–time graph

Speed–time graphs

A speed–time graph shows the speed of a moving object at every instant during its journey. The speed–time graph for Vijay's trip on page 76 would then look like the graph on the right. A steady speed is now shown by a straight horizontal line.

Again, the sudden changes of speed shown on this graph are not realistic. It would take time for the speed of a moving object to change. A more realistic speed–time graph would have smoother, more gradual changes from one speed to another.

The second speed–time graph on the right is for a tennis ball being dropped to the ground. It has no horizontal sections, so the speed of the ball is changing all the time. The constant slope of the speed–time graph shows that the speed is changing at a steady rate. It has steady (or uniform) acceleration downwards.

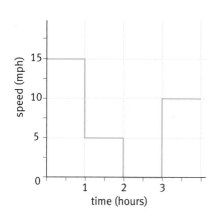

Speed–time graph for Vijay's cycle trip

Questions

2 What is the meaning of:
   i   a horizontal section
   ii a section with a steady upward slope
   iii a section with a steady downward slope

   a   on a distance–time graph?

   b   on a speed–time graph?

3 Roberta, an athlete, trains by jogging 20 m at a steady speed, then sprinting 20 m. She repeats this five times. Sketch a speed–time graph of her motion during a training session.

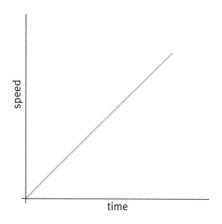

Speed–time graph for a falling tennis ball

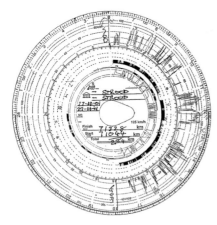

A tachograph trace – a speed–time graph of the vehicle's motion during a 24-hour period

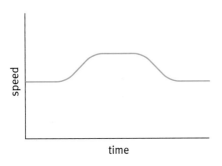

A speed–time graph for part of a lorry journey.

Key words

distance–time graph
velocity
velocity–time graph

A velocity–time graph. The negative velocity means that Karl on his skateboard is travelling in the opposite direction to his original motion.

Tachographs

According to EU regulations, lorry drivers are only allowed to drive for 9 hours per day. And they must take a break of at least 45 minutes every 4.5 hours. Lorries are also subject to speed limits on the road. So haulage companies have to keep a check on what their drivers are doing. They do this by installing a tachograph on each of their lorries. The tachograph monitors the lorry's distance and speed. It draws a graph of the lorry's speed against time. A tachograph trace is shown on the left.

An enlarged (and simplified) section of a tachograph trace might look like the speed–time graph below left. In the first section, the graph is horizontal. This indicates a constant speed. It then slopes upwards: the speed is increasing. The lorry travels for a while at a higher steady speed (another horizontal section). It then slows down again to its original speed.

Velocity–time graphs

Look at the graph below. It shows the motion of Karl's skateboard up and then down a slope. At first Karl was travelling at 10 mph. He gradually slows down until his speed is zero. But then the line of the graph keeps on going down. The speed becomes negative. This may seem strange, but it is used to show that the skateboard's direction of motion has changed. Karl is now travelling in the opposite direction.

When you want to talk about speed in a certain direction, you use the term **velocity**. Velocity simply means speed in a certain direction.

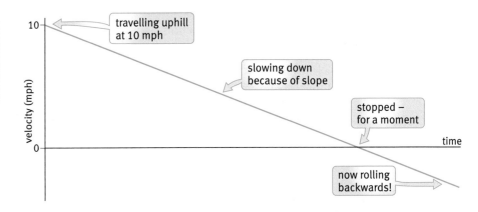

Questions

4 Sketch a distance–time graph for the part of the lorry's journey shown in the speed–time graph above left.

5 How would the speed–time graph of Karl on his skateboard be different from the velocity–time graph? Sketch a speed–time graph.

6 Think about the motion of a ball thrown upwards. Its speed gets steadily less on the way up and increases steadily again on the way down. Draw its

  a speed–time graph

  ⓑ velocity–time graph

from the moment it leaves your hand until the moment it lands back in your hand.

H Force, interaction, and momentum

Forces and motion

The key idea for explaining motion is **force**. If you know the forces acting on an object, you can predict how it will move. Look in more detail at what happens when two objects interact. Think about the three situations shown in the diagrams below:

Find out about:
▶ momentum
▶ the link between change of momentum, force, and time

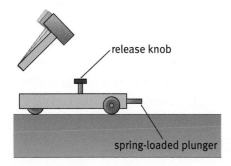

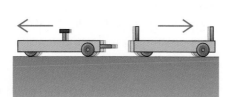

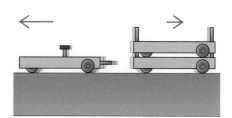

1 This trolley has a spring-loaded plunger. When the spring is released, by tapping the knob on top, the plunger springs out. But the trolley does not move. It has nothing to interact with. It cannot exert a force. So it does not experience a force in return.

2 If you put a second trolley in front of the first one and then release the spring, both move, in opposite directions. The interaction causes two forces, one on each trolley. If the trolleys are identical, they both move with the same speed.

3 Here one trolley is twice as heavy as the other. When the spring is released, both move. But the heavier one has only half the speed of the lighter one.

When there is no interaction, there is no force on the trolley. So it does not move. When there is an interaction, both objects move. If the objects have different masses, the heavy one moves more slowly than the light one. In fact, the number you get if you calculate 'mass × speed' is the same for both. This seems to be an important quantity, so it is called the **momentum** of the object. Because the direction matters, momentum is defined by the equation

$$\underset{\substack{\text{(kilogram metre} \\ \text{per second, kg m/s)}}}{\text{momentum}} = \underset{\substack{\text{(kilogram, kg)}}}{\text{mass}} \times \underset{\substack{\text{(metre per} \\ \text{second, m/s)}}}{\text{velocity}}$$

So, if an object is moving in one direction, its momentum is positive. And if it is moving in the other direction, its momentum is negative. You can choose which direction to call 'positive' in any situation.

Questions

1 What would you expect to happen if you carried out the exploding trolley investigation above with a stack of three trolleys on the right? Explain why.

2 What is the momentum of:

a a skier of mass 50 kg moving at 5 m/s?

b a netball of mass 0.5 kg moving at 3 m/s?

c a whale of mass 5000 kg swimming at 2 m/s?

3 A snooker ball has a momentum of 1 kg m/s just before it hits a cushion head-on. It bounces straight back with the same speed. What is its momentum now? Has its momentum changed?

Taking a free kick. The interaction between the footballer's foot and the ball causes a change of momentum.

Force and change of momentum

When a footballer takes a free kick, there is an interaction between his foot and the ball. His foot exerts a force on the ball. And the ball exerts a force on his foot. (That is why it can hurt to kick a ball with bare feet!)

This force lasts for only a very short time, the time for which the foot and the ball are actually in contact. After that, the player's foot can no longer affect the motion of the ball. The ball is on its own. So it is wrong to think of a kick 'giving the ball some force' or 'putting some force into the ball'. But the kick does give the ball some momentum. It causes a *change in momentum.*

Causing a change of momentum

Imagine pushing an object hard enough to make it start moving. If you keep pushing, it will get faster and faster. Its momentum is increasing. The change of momentum depends on two things:

▶ the size of the force you push with

▶ the time for which you keep pushing

This can be written in the form of an equation:

change of momentum = force × time for which it acts
(kilogram metre per second, kg m/s)   (newton, N)   (second, s)

All this is consistent with what you have seen in the interactions between two trolleys on page 79. After the interaction, the momentum of each of the moving objects is the same size. But in any interaction, the two objects involved always experience equal and opposite forces. These forces last for exactly the same length of time: the duration of the interaction. So 'force × time for which it acts' is also the same for both objects.

Questions

4 When a force makes an object move, which two factors determine the change of momentum of the object?

5 Which of the following will cause:

  i the largest change in momentum?

  ii the smallest change of momentum?

  a a force of 40 N acting for 3 s

  b a force of 200 N acting for 0.5 s

  c a force of 3 N acting for 50 s?

Example

A football has a mass of around 1 kg. A free kick gives it a speed of 20 m/s. What is its momentum?

The football's momentum is given by the equation

momentum = mass × velocity
= 1 kg × 20 m/s
= 20 kg m/s

As it started with speed zero (when it was not moving), the ball's change of momentum during the kick is 20 kg m/s.

Using high-speed photography, it is possible to measure the contact time when a football is kicked. It is around 0.05 s (or one-twentieth of a second).

Estimate the force exerted on the ball during the kick.

Use the equation

change of momentum = force × time for which it acts
20 kg m/s = force × 0.05 s

Dividing both sides by 0.05 s, you get

$$\frac{20 \text{ kg m/s}}{0.05 \text{ s}} = \text{force}$$

So the force on the ball during the kick is 400 N. This is equal to the weight of a 40 kg object. No wonder it can hurt your foot! In fact, this is the *average* force during the kick. The maximum force will be even bigger.

Conservation of momentum

When there is an interaction between two objects, the change of momentum of one is equal in size to the change of momentum of the other but is opposite in direction. Another way to say this is:

▶ When two objects interact, the total change in momentum of the two objects (taking direction into account) is zero.

So the total momentum of the two objects is the same after the interaction as it was before.

This is true for any interaction. So it is a useful and important result. Scientists call it the principle of conservation of momentum.

You can use the idea of conservation of momentum to predict the speed of objects after an interaction. Look at the two skaters in the diagrams on the right. Zelda pushes on Jake's hands, and both move apart. Jake's mass is 60 kg and he is moving at 2 m/s. So his momentum is 120 kg m/s. You therefore know that Zelda's momentum must also be 120 kg m/s, in the opposite direction. As her mass is 40 kg, you can work out that she is moving at 3 m/s:

$$\text{momentum} = \text{mass} \times \text{velocity}$$

$$\text{velocity} = \frac{\text{momentum}}{\text{mass}}$$

$$= \frac{120\,\text{kg m/s}}{40\,\text{kg}} = 3\,\text{m/s}$$

Notice that the lighter person has the higher speed. This is just like the interacting trolleys on page 79.

The interactions you have been looking at so far are 'explosions', where two objects push apart. Collisions are another type of interaction. In any collision, momentum is also conserved.

Before

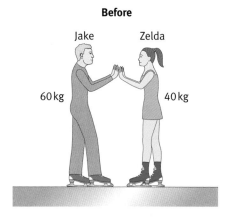

After

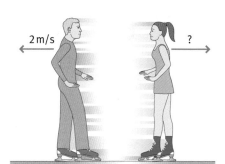

Jake pushes Zelda and they both move apart.

Key words
momentum

Questions

6 Look at this sequence of high-speed photographs of a tennis shot.

a If the frames are 0.01 s apart, estimate how many seconds the interaction between the ball and the racket lasts.

b The mass of a tennis ball is 0.06 kg. If it is moving at 12 m/s after this shot, estimate the force exerted on the ball by the racket.

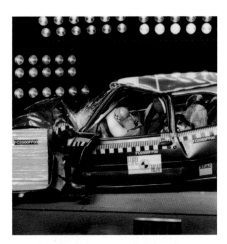

This driver, Hybrid 111, has experienced many crashes. With a steel skeleton and rubber skin it is packed with sensing equipment to record forces on different areas, like the head, the chest, and the neck. Each of these dummies costs more than £100,000 to build.

Questions

1 When you jump down from a wall or ledge, it is almost automatic to bend your knees as you land. Use the ideas on this page to explain why this reduces the risk of injury.

2 In railway stations, there are buffers at the end of the track. These are a safety measure – designed to stop the train if its brakes failed. Use the ideas on this page to explain how buffers would reduce the forces acting on the train and on the passengers, in the event of an accident of this sort.

Car safety

Cars today are much safer to travel in than cars ten or twenty years ago. As a result of crash tests like the one shown on the left, designs have changed and are still changing.

If a car is travelling at 70 mph, the driver and passengers are also travelling at that speed. If the car comes to a very sudden stop, owing to a collision, the occupants will experience a very sudden change in their momentum. This could cause serious injury.

Crumple zones

Look at the diagram below. Which car would be safer in a collision? The answer may not be so obvious. You need to think about the change of momentum during the collision and the time the collision lasts.

a

b

Would you be safer in car **a** or car **b**?

The momentum of a moving car depends on

▶ its mass

▶ its velocity

In a collision, the car is suddenly brought to a stop. Its momentum is then zero. The size of the force exerted on the car during the collision depends on the time the collision lasts:

change of momentum = force × time for which it acts

The bigger the time, the smaller the force – for the same change of momentum. This is why cars are fitted with front and rear crumple zones, with a rigid box in the middle. They are designed to crumple gradually in a collision. This makes the duration of the collision (the time it lasts) longer. This then makes the force exerted on the car less.

The passengers inside the car also experience a sudden change of momentum. A force exerted on their bodies (by whatever they come into contact with) causes this change. The longer it takes to change the passengers' speed to zero, the smaller the force they experience.

Seat belts and air bags

Some people think that seat belts work by stopping you moving in a crash. In fact, to work, a seat belt actually has to stretch. Seat belts work on the same principle as crumple zones. They make the change of momentum take longer. So the force that causes the change is less.

With a seat belt, the top half of your body will still move forward, and you may hit yourself against parts of the car. Air bags can help to cushion the impact. Again, they reduce your momentum more slowly so that the force you experience is less.

Could you save yourself?

Some people think they could survive a car accident without a seat belt, especially if they are travelling in the back seats. The diagram below shows the position of a dummy driver at different times after an impact. The car was originally travelling at just 30 mph (or roughly 14 m/s). Without a seat belt, the dummy hits the steering wheel and windscreen about 0.07 s after the impact. Back-seat passengers would hit the back of the front seats at roughly the same time. As your reaction time is typically about 0.14 s, this would all happen before you even have time to react. Even if you could react in time, the force needed to change your speed from 14 m/s to zero in 0.1 s is larger than your arms or legs could possibly exert.

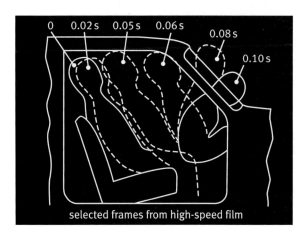

selected frames from high-speed film

The position of a dummy driver at a series of instants after a crash.

How seatbelts work. Notice how the seatbelt stretches during the collision. This 'spreads' the change of the driver's momentum over a longer period, making the force he experiences smaller.

Questions

3 Using the diagrams on the right, estimate how long it takes for the seat belt to bring the driver's body to a stop. If his mass is 70 kg, what is the average force that the belt has to exert to do this?

4 If the belt were made of material that did not stretch as much, would the force exerted on the driver be larger or smaller? Explain why. What if it were made of a material that stretched more?

Find out about:
- the laws (or rules) that apply to every example of motion
- how a resultant force is needed to change an object's motion

1 Smooth floor

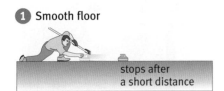

stops after a short distance

Imagine pushing this curling stone across a smooth floor. It will keep going after it leaves your hand, because of the momentum you have given it during the interaction with your hand. But it immediately begins to slow down because of friction, and soon it will stop.

2 Ice

goes further before stopping

Now think what would happen if you gave the same stone exactly the same push, but this time on ice. It would not slow down as quickly. But it would slow down all the same, because there is still some friction. Eventually it would stop.

3 'Perfect' ice

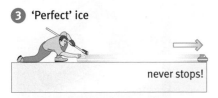

never stops!

Now imagine 'perfect' ice, so slippery that there is no friction force between it and the stone. If you were able to give the stone the same push as before, it would not slow down after it left your hand. There is no friction force, so it just keeps on going, at the same speed. for ever.

Key words
driving force
counter-force

ᴊ Laws of motion

In everyday situations, there are always several forces acting on each object involved. To find out what will happen to the object, you need to find out what the combined effect of these forces is. To do this, you add the forces, taking their directions into account. This gives you the resultant force acting on the object (see page 73). You then apply the following laws (or rules) of motion:

- **Law 1:** If the resultant force acting on an object is zero, the momentum of the object does not change.

- **Law 2:** If there is a resultant force acting on an object, the momentum of the object will change. The change of momentum is given by (change of momentum = resultant force × time for which it acts) and is in the same direction as the resultant force.

These laws are completely general. They apply to every example of motion. That is why they are so useful.

When the resultant force is zero

Stationary objects are one example of law 1. If an object is stationary, its momentum is not changing. It is zero all the time. The resultant force acting on it is also zero. The forces acting on the object are balanced. They cancel each other out.

Now think about about an object travelling with constant velocity. Its momentum is not changing. The first law above says that the resultant force acting on it is zero. But you might be thinking that there must be a resultant force in the direction the object is going, to keep it moving. *In fact, this is not correct.* To see why, look at the diagrams on the left.

In the real world, there is *always* friction. So a **driving force** is needed to keep an object moving. But this just has to balance the friction force so that the resultant force *is* zero. If the driving force were bigger than the friction force. The object would not move at a steady speed. There would now be a resultant force. Its momentum would increase and it would speed up. Look now at how this works in an everyday situation: riding a bicycle.

Forces acting on a cyclist

When you press on the pedals of a bike, the chain makes the back wheel turn. The tyre pushes back along the ground. The other force in this interaction pair is the force exerted by the ground on the tyre. This pushes the bike forward. It is therefore called the driving force. As you move, air resistance and friction at the axles cause a **counter-force**. This is in the opposite direction to your motion.

1 When you are starting off, the counter-force is very small. Your driving force is bigger. So you move forward, and your speed increases.

counter-force   driving force

2 As you go faster, the air resistance force on you gets bigger. So the counter-force increases. You are still getting faster, but not as quickly as before.

counter-force   driving force

3 Eventually you reach a speed where the counter-force exactly balances your driving force. Now your speed stops increasing. You carry on travelling, at a steady speed.

counter-force   driving force

A cyclist riding along at a steady speed. Is the resultant force on the cyclist zero? Or is there a resultant force forwards, in the direction she is going?

Questions

1 List three examples from everyday life of a situation where the resultant force on an object is zero. Explain how these are in agreement with the first law of motion on page 84.

2 List three examples from everyday life of a situation where there is a resultant force acting on an object. Explain how these are in agreement with the second law of motion on page 84.

3 Draw a fourth diagram in the series on this page to show the forces acting when the cyclist stops pedalling and freewheels. Write a caption, like those for the first three diagrams, to explain the motion.

Find out about:

▶ how to calculate the work done by a force

▶ the link between work done on an object and the energy transferred

▶ how to use energy ideas to predict the motion of objects

Pushing a car along is hard work!

K Work and energy

So far, you have seen how the ideas of force and momentum explain motion. However, it is sometimes hard to use these ideas to make an exact prediction, even though they apply to all situations. For example, to predict the speed something will reach when it slides down a slope, it is easier to use energy ideas. But first you need to make the connection between force and energy. The link is the idea of **work.**

Doing work

Imagine that you are out in your car and you break down. Luckily, there is a garage just down the road. You ask your passenger to steer the car while you push it to the garage. To do this, you have to transfer energy from your store of chemical energy (in your muscles) to the car. We say that you have to do work.

If the garage is a long way down the road, you are going to have to do more work than if it is nearby. You will also do more work the harder the car is to push. So the amount of work depends on

▶ the force you have to exert

▶ the distance moved in the direction of the force

The amount of work done by a force is defined by this equation;

work done by a force = force × distance moved in direction of force
(joule, J)   (newton, N)   (metre, m)

Example

Calculate how much work is needed to push a car 50 m along a road.

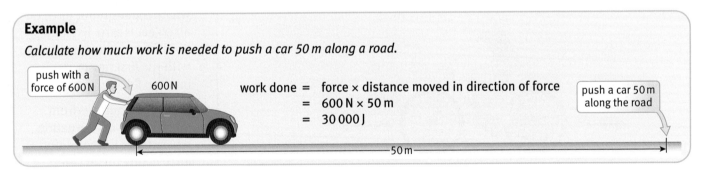

push with a force of 600 N   600 N

work done = force × distance moved in direction of force
= 600 N × 50 m
= 30 000 J

push a car 50 m along the road

50 m

The amount of work that you do is equal to the amount of energy you transfer:

amount of work done = amount of energy transferred

Work, like energy, is measured in joules (J).

Lifting things: changing their gravitational potential energy

Lifting luggage into the boot of the car also involves doing work. You are transferring energy from your body's store of chemical energy. The **gravitational potential energy** of the luggage increases. The increase is equal to the amount of work you have done.

Suppose you have a suitcase that weighs 300 N. To lift it up, you have to exert an upward force of 300 N. If you lift it 1 metre into the boot of the car, then

$$\text{work done} = \text{force} \times \text{distance moved in the direction of the force}$$
$$= 300\,\text{N} \times 1\,\text{m}$$
$$= 300\,\text{J}$$

So the suitcase gains 300 J of gravitational potential energy. In general, when anything is lifted up, you can calculate its change in gravitational potential energy from the equation

$$\underset{\text{(joule, J)}}{\text{gravitational potential energy}} = \underset{\text{(newton, N)}}{\text{weight}} \times \underset{\text{(metre, m)}}{\text{vertical height difference}}$$

Notice that it is only the vertical height difference that matters. If you slide a suitcase up a ramp, the gain in gravitational potential energy is the same as if you lift it vertically. However, you would have to do more work because some energy is wasted on heating the ramp and suitcase, owing to the friction between them.

Making things speed up: changing their kinetic energy

Imagine pushing a well-oiled supermarket trolley along a level floor. As you push, you are doing work. The trolley keeps speeding up. Its speed increases as long as you keep pushing. You are transferring energy from your body's store of chemical energy to the trolley, where it is stored as kinetic energy. If the trolley is absolutely smooth running, the amount of work you do pushing it is equal to the change in the trolley's kinetic energy.

Doing work by lifting: increasing gravitational potential energy

> **Example**
>
> *Calculate the change in kinetic energy of a trolley pushed with a force of 6 N over a distance of 5 m (assume there are no frictional forces acting).*
>
> $$\text{change in kinetic energy of trolley} = \text{work done by pushing force}$$
> $$= \text{force} \times \text{distance}$$
> $$= 6\,\text{N} \times 5\,\text{m}$$
> $$= 30\,\text{J}$$
>
>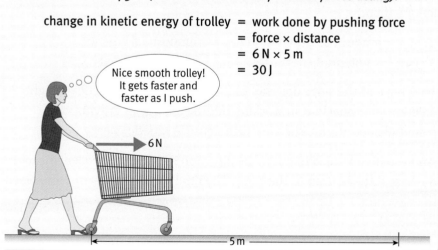

A real trolley will always have some friction, so its change in kinetic energy will be less than this. Some work is wasted in causing unwanted heating (and sound).

Questions

1 It takes a force of 1200 N to push a large car along the road. How much work would you have to do to push it 40 m?

2 The equation for work done by a force (page 86) is similar to the equation for the change of momentum caused by a force (page 80). But it has one important difference. What is it?

3 If your mass is 40 kg, then your weight is roughly 400 N. How much work do you have to do each time you go upstairs – a vertical height gain of 2.5 m?

4 A mother is pushing a child along in a buggy. She is doing work. So the amount of energy stored in her muscles is getting less. Where is this energy being transferred to? (Careful! The buggy is going at a steady speed.)

The equation for calculating the **kinetic energy** of a moving object is:

$$\text{kinetic energy}_{\text{(joule, J)}} = \frac{1}{2} \times \text{mass}_{\text{(kilogram, kg)}} \times \text{(velocity)}^2_{\text{(metre per second, m/s)}^2}$$

Notice that the amount of kinetic energy depends on the velocity squared. So small changes in velocity mean quite big changes in kinetic energy.

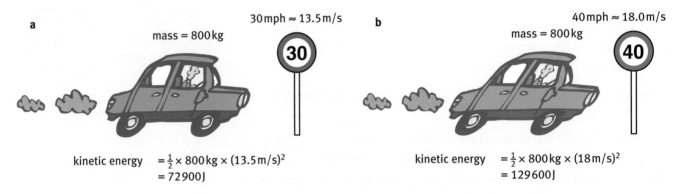

a mass = 800 kg 30 mph ≈ 13.5 m/s

kinetic energy $= \frac{1}{2} \times 800\,\text{kg} \times (13.5\,\text{m/s})^2$
$= 72\,900\,\text{J}$

b mass = 800 kg 40 mph ≈ 18.0 m/s

kinetic energy $= \frac{1}{2} \times 800\,\text{kg} \times (18\,\text{m/s})^2$
$= 129\,600\,\text{J}$

A car travelling at 40 mph (**b**) has nearly twice as much kinetic energy as the same car at 30 mph (**a**). This explains why there is a much greater risk of injury to pedestrians at at speeds greater than 30 mph.

All the fun of the fair

When a roller coaster runs down a slope, it

▶ loses gravitational potential energy

▶ gains kinetic energy

If the slope has a complicated shape, it would be very difficult, maybe even impossible, to work out how fast the roller coaster is going at the bottom of the slope using the ideas of force and momentum. It is easier to use the principle of **conservation of energy**. If friction is small enough to ignore, then:

$$\text{amount of gravitational potential energy lost} = \text{amount of kinetic energy gained}$$

The shape of the slope does not matter at all. Only the vertical height difference is important in working out the change in gravitational potential energy – and hence the increase in kinetic energy.

And you do not need to know the direction in which the roller coaster is moving at the bottom of the slope. The final speed is the same, whatever the direction.

The 'Oblivion' vertical roller coaster ride at Alton Towers has one section where the drop is vertical!

Key words

work
gravitational potential energy
kinetic energy
conservation of energy

To see how this works out in practice, look at the following example:

Example

Calculate the speed of the roller coaster shown below at the bottom of the ride (assuming there are no friction forces).

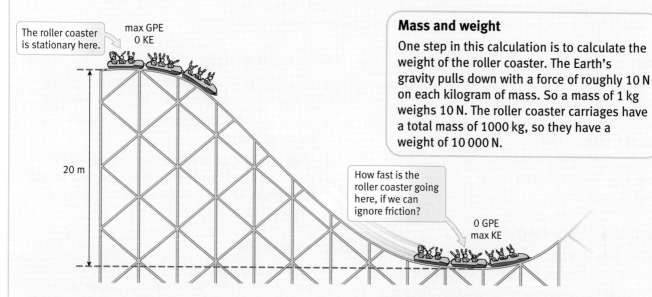

The roller coaster is stationary here.

max GPE
0 KE

20 m

How fast is the roller coaster going here, if we can ignore friction?

0 GPE
max KE

Mass and weight

One step in this calculation is to calculate the weight of the roller coaster. The Earth's gravity pulls down with a force of roughly 10 N on each kilogram of mass. So a mass of 1 kg weighs 10 N. The roller coaster carriages have a total mass of 1000 kg, so they have a weight of 10 000 N.

The simplest way to solve this problem is to use energy ideas. As there is no friction, the gravitational potential energy that the roller coaster loses as it goes down the slope is equal to the kinetic energy it gains:

1   loss of gravitational potential energy   = weight × vertical height change
                                                      = 10 000 N × 20 m
                                                      = 200 000 J

2   loss of gravitational potential energy   = gain in kinetic energy

3 So   gain in kinetic energy   = 200 000 J

4 But   gain in kinetic energy   = $\frac{1}{2}$ × mass × (velocity)^2

   So $\frac{1}{2}$ × mass × (velocity)^2 = 200 000 J

   Multiply both sides by 2:   mass × (velocity)^2 = 400 000 J

   or   1000 kg × (velocity)^2 = 400 000 J

   Divide both sides by 1000 kg: (velocity)^2 = $\dfrac{400\,000\ \text{J}}{1000\ \text{kg}}$

                                                 = 400 (m/s)^2

   Take the square root of both sides: velocity   = 20 m/s

Questions

5 A ten-pin bowling ball has a mass of 4 kg. It is moving at 8 m/s. How much kinetic energy does it have?

6 Which of the following has more kinetic energy?

   a a car of mass 500 kg travelling at 20 m/s

   b a car of mass 1000 kg travelling at 10 m/s

7 Repeat the calculation above for a roller coaster that is only half as heavy (weight 5000 N). What do you notice about its speed at the bottom? How would you explain this?

P4 Explaining motion

Summary

This module has all been about explaining how and why things move as they do. A few laws of motion can account for all the kinds of motion you see around you.

Interactions and forces

- When two objects interact (by contact or action at a distance), both experience a force.
- These two forces are equal in size but opposite in direction. Each acts on a different object.
- Vehicles (and people) move by pushing back on something. This interaction causes a forward force to act on them.

Friction and normal reaction

- Friction is an interaction between two objects that are sliding (or tending to slide) past each other.
- The friction force matches the applied force that is making the objects slide – up to a limit.
- When an object sits on a surface, it distorts it slightly. The 'springiness' of the surface then causes a reaction force on the object – matching its downward push on the surface.

Describing motion

- The average speed of a moving object is

$$\frac{\text{distance}}{\text{time taken}}$$

- The instantaneous speed of a moving object is its speed at a particular instant. To estimate this, you measure its average speed over a very short distance (or time).
- Velocity means speed in a particular direction.
- Distance–time, speed–time, and velocity–time graphs are useful for summarizing and analysing the motion of an object.

Forces and motion

- When a force acts on an object, it causes a change in its momentum. Momentum is 'mass × velocity'.
- The change of momentum is equal to 'force × time for which it acts'.
- Many vehicle safety features work by making the time of an event (such as a collision) longer, so that the average force is less, for the same change of momentum.

Laws of motion

- If the resultant force acting on an object is zero, the momentum of the object does not change. If it is stationary, it does not move. If it is moving, it will keep moving at a constant speed in a straight line.
- If the resultant force acting on an object is not zero, this will cause a change in its momentum, in the direction of the resultant force.

Work and energy

- Work is done when a force makes an object move. The amount of work is 'force × distance'.
- When something does work, its energy decreases by that amount. If work is done on it, its energy increases by that amount.
- Doing work on an object can increase its gravitational potential energy (by lifting it up) or its kinetic energy (by making it move faster).
- Change in gravitational potential energy is 'weight × vertical height difference'. Change in kinetic energy is '½ × mass × velocity²'
- When an object drops to a lower level, it loses gravitational potential energy. If friction can be ignored, it gains the same amount of kinetic energy.
  This can be used to work out its speed at the bottom. H

Questions

1 Think about the following situations:

   i Amjad on his skateboard, throwing a heavy ball to his friend (main objects to consider: Amjad + skateboard; the ball).

   ii A furniture remover trying to pull a piano across the floor – but it will not move (main objects to consider: the furniture remover; the piano; the floor).

   iii A hanging basket of flowers outside a café (main objects to consider: the basket; the chain it is hanging from).

   For each of them:

   a Sketch a diagram (looking at it from the side).

   b Then sketch separate diagrams of the main objects in the situation (these are listed for each).

   c On these separate diagrams, draw arrows to show the forces acting on that object. Use the length of the arrow to show how big each force is.

   d Write a label beside each arrow to show what the force is.

2 Imagine what it would be like to wake up one morning and discover that friction had disappeared. Write a short story about what it would be like to live in a friction-free world.

3 A tin of beans on a kitchen shelf is not falling, even though gravity is still acting on it. The shelf exerts an upward force, which balances the force of gravity. Explain in a short paragraph how it is possible for a shelf to exert a force. Draw a sketch diagram if it helps your explanation.

4 **a** The winner of a 50 m swimming event completes the distance in 80 s. What is his average speed?

   b How far could Leonie cycle in 10 minutes if her average speed is 8 m/s?

   c The average speed of a bus in city traffic is 15 m/s. How much time should the timetable allow for the bus to cover a 6 km route?

5 What is the momentum of:

   a a hockey ball of mass 0.4 kg moving at 5 m/s?

   b a jogger of mass 55 kg, running at 4 m/s?

   c a van of mass 10 000 kg, travelling at 15 m/s?

   d a car ferry of mass 20 000 000 kg, moving at 0.5 m/s?

6 Which of the following would cause the biggest change of momentum? And which would cause the smallest?

   a a force of 35 N acting for 4 s

   b a force of 3 N acting for 50 s

   c a force of 1500 N acting for 0.1 s

   d a force of 8 N acting for 20 s

   Explain how you worked this out.

7 Your uncle, who last studied science many years ago at school, thinks that you obviously need a force to keep something moving. As he says, 'If you stop pushing something, it stops moving'. Write him a note, explaining why he is wrong. Include plenty of examples, to convince him of your argument.

8 A weightlifter raises a bar of mass 50 kg until it is above his head – a total height gain of 2.2 m. How much gravitational potential energy has it gained? How much more work must he do to hold it there for 5 s?

9 A packet of mass 2 kg falls from the upstairs window of a flat, 45 m above the ground.

   a How much gravitational potential energy has the packet lost just before it hits the ground?

   b If we ignore the effect of air resistance, how much kinetic energy does the packet have just before it hits the ground?

   c With what speed does it hit the ground?

   d Would the speed be different for a similar packet of mass 5 kg?

   e If the packet had slid down a smooth chute instead of falling, how would its speed at the bottom compare?

Why study growth and development?

How does a human embryo develop? What makes cells with the same genes develop differently? Exploring questions like these is part of the fast-moving world of modern biology.

The science

Sex cells carry the genetic information to make a new individual. After fertilization, the egg cell divides many times to form an embryo. Cells become specialized because of the different proteins they make. DNA is the chemical that genes are made of. It has a unique structure which determines the proteins a cell makes. There are differences in how plants and animals grow, and we use these differences to grow cloned plants.

Biology in action

Genetic technologies are at the cutting edge of modern science. Our expanding knowledge of how human beings grow and develop has the potential to offer great benefits for this and future generations. For example, research into how cell growth is controlled could be crucial in combating cancer.

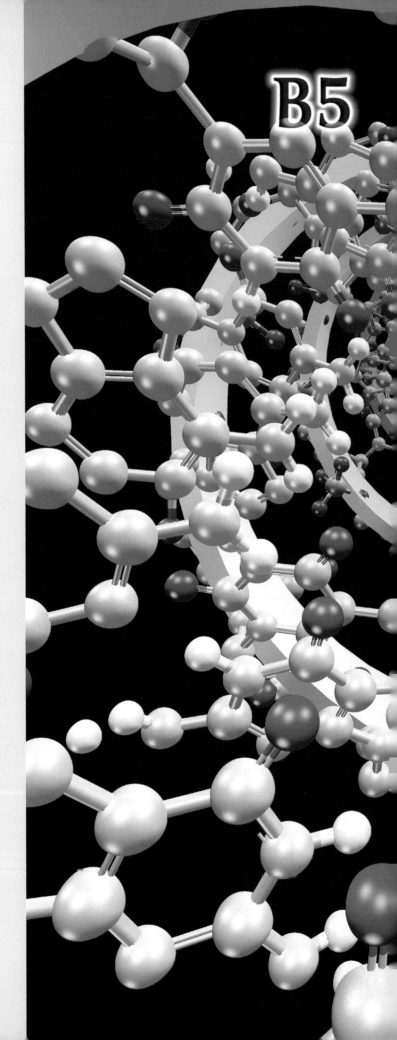

B5

Growth and development

Find out about:

- the structure of DNA, and how it controls the proteins a cell makes
- how cells divide to make sex cells and to make new body cells
- how cells become specialized
- the differences between plant and animal growth

A Growing and changing

Find out about:
- different cells, tissues, and organs
- growing up

You began life as a single cell. By the time you were born you were made of millions of cells. You probably weighed about 3 or 4 kilogrammes. Now you probably weigh over 50 kilogrammes. Not only have you grown, but you have changed in many ways. In other words, you have developed.

Since you were born, your **development** has been gradual. In some plants and animals, development involves big changes. For example, the young and the adults in these photographs look very different:

Key words

development

Life cycles

Until 1668 people thought that maggots just appeared in decaying meat. The scientist Francesco Redi observed flies on the meat with maggots. He did experiments to find out if they were linked. He put pieces of meat in pots and covered some of them with gauze. He left the others uncovered. Maggots appeared only on the meat in the uncovered pots. Where flies could not get to the meat, there were no maggots.

Redi eventually worked out the whole life cycle of the fly:

- Flies lay eggs.

- Eggs hatch into maggots.

- Maggots change into pupae. Inside the pupa, the tissues reorganize into a fly.

Questions

1 Match the young and the adults in pictures A to F.

② Which animal in the pictures has a life cycle most like a fly?

Building blocks

Like you, the plants and animals in the pictures opposite are multicellular. Your body has more than 300 different kinds of cell. Each cell is **specialized** to do a particular job.

Tissues and organs

All newly formed human cells look much the same. Then they develop into groups of specialized cells called **tissues**.

Plant cells are different from animal cells, but they are specialized too. Plant cells have walls, and some have spaces called vacuoles.

As an animal embryo or plant grows, groups of tissues arrange themselves into **organs**, for example the heart and brain in humans, and roots, leaves, and flowers in plants.

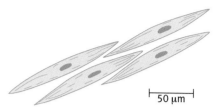

Muscle cells contract and relax to cause movement.

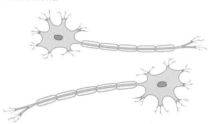

Not to scale

Nerve cells carry nerve impulses.

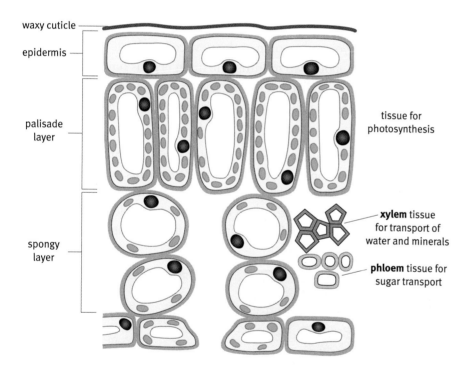

Different tissues in a leaf – a plant organ.

Questions

3 What does *specialized* mean?

4 What is the job of these tissues:
  a muscle?
  b xylem?
  c phloem?

5 Name three plant organs.

6 Explain the difference between a tissue and an organ.

Key words

development
specialized
tissues
organs
xylem
phloem

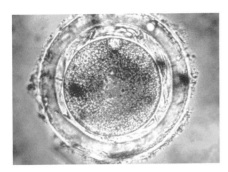

A fertilized human egg cell

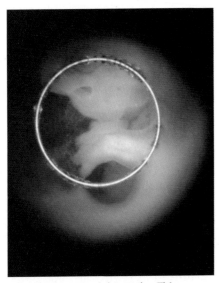

A human fetus at eight weeks. This photograph was taken from inside the mother's uterus. The fetus is about 2.5 cm long.

Key words

zygote
fetus
embryonic stem cells
meristem cells

Questions

7 What is a zygote?

8 When does a human embryo become a fetus?

From single cell to adult

All the cells in your body come from just one original cell – a fertilized egg cell or **zygote**.

So the zygote must contain instructions for making all the different types of cells in your body, for example muscle cells, bone cells, and blood cells. It also has the information to make sure that each type of cell develops in the right place and at the right time. This information is in your DNA – the chemical that your genes are made of.

In humans:

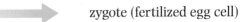

fertilization

sperm + egg cell   zygote (fertilized egg cell)

The growing baby

During the first week of growth, the zygote develops into a ball of about 100 cells. The nucleus of each cell contains an exact copy of the original DNA. As the embryo grows, some of the new cells become specialized and form tissues. After about two months, the main organs have formed and the developing baby is called a **fetus**. A six-day embryo is made of about 50 cells. Adults contain about 10^{14} cells, each with the DNA faithfully copied.

When the embryo is a ball of cells, it occasionally splits into two. A separate embryo develops from each section. When this happens, identical twins are produced. They are clones of each other. This shows that there are cells in the early embryo that can develop into complete individuals. These are **embryonic stem cells.** You can read more about these cells in Section H.

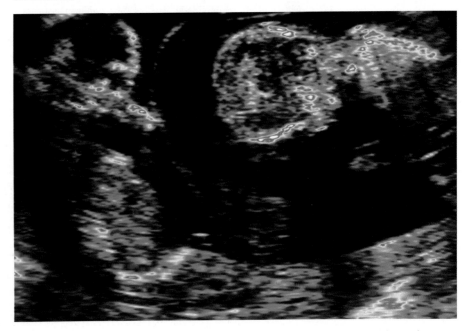

This ultrasound scan shows that twins are expected. Both of the babies' heads can be seen.

Growth patterns

For living things to grow bigger, some of their cells must divide to make new cells. You will probably stop growing taller by the time you are about 18–20 years old.

Flowering plants continue to grow throughout their lives.

▶ Their stems grow taller.

▶ Their roots grow longer.

▶ To hold themselves upright, most increase in girth or have some other means of support.

Plants increase in length by making new cells at the tips of both shoots and roots. They also have rings of dividing cells in their stems and roots to increase their girth. These dividing cells are called **meristem cells**.

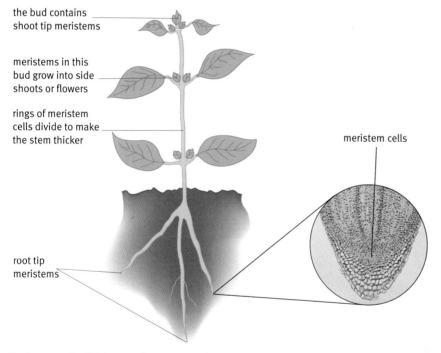

the bud contains shoot tip meristems

meristems in this bud grow into side shoots or flowers

rings of meristem cells divide to make the stem thicker

meristem cells

root tip meristems

Meristem cells divide to make stems and roots longer and make the stem thicker. On the right is a root tip meristem (photographed through a microscope).

This giant sequoia tree is over 2000 years old, 83 m tall, and 26 m in girth (circumference). It is known as 'General Sherman' and is officially the largest giant sequoia tree and the largest living thing on Earth.

Questions

9 Why is it important that all living things have cells that can divide?

10 Name the type of cell in a plant that can divide.

11 Explain how plants:
   a grow taller
   b grow longer roots
   c grow thicker in girth

Honeysuckle twines its stem around other plants for support.

97

Find out about:

▶ why plants are so good at repairing damage

B Growing plants

Cells in your body divide when you are growing. If you cut yourself, cells can also divide to repair your body. But your body can make only small repairs. Many plants and some animals can replace whole organs.

Why can plants grow back?

Plant meristem cells are **unspecialized**. Plants keep some meristem cells all through their lives. These are spare back-up cells that can divide to make any kind of cell the body needs. So plants can regrow whole organs, such as leaves, if they are damaged.

How do newts grow?

Animals also have spare back-up cells called **stem cells**. These cells divide, grow, and develop into any kind of cell the body needs.

Animal growth

Newts' stem cells stay unspecialized throughout their life. So newts can grow new legs if they need to – or even an eye.

The stem cells in adult humans are not as useful, because they are already specialized. For example, the stem cells in your skin can only develop into skin cells.

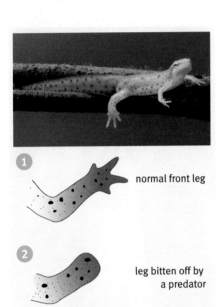

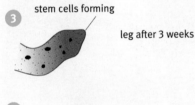
① normal front leg

② leg bitten off by a predator

③ stem cells forming
leg after 3 weeks

④ leg after 4 weeks

⑤ leg after 6 weeks

⑥ leg after 10 weeks

If a newt's limb is bitten off by a predator, it can grow a replacement. Most animals can only make small repairs to their body.

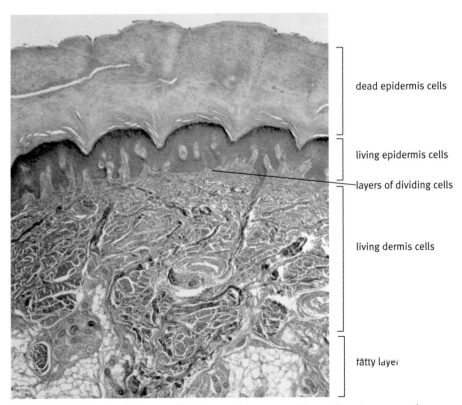

dead epidermis cells

living epidermis cells

layers of dividing cells

living dermis cells

fatty layer

A cross-section through human skin. Some of the stem cells continue to grow and divide. Others replace skin cells at a wound or those that wear off at the surface. (×36)

You replace millions of skin cells every day. Most of the dust in your home is worn-off skin cells. Other tissues in your body need a constant supply of new cells. For example, bone marrow contains stem cells to make new blood cells.

Using meristems to make more plants

Gardeners use meristems when they grow new plants by taking **cuttings.** Cuttings are just shoots or leaves cut from a plant. In the right conditions they develop roots and grow into new plants.

Some cuttings grow new roots when you put them in water or compost. Others grow better when you dip the cut ends in **rooting powder** before you plant them. Rooting powder contains plant hormones called **auxins.** Auxins make the new cells produced by the shoot meristem develop into roots.

By taking cuttings, gardeners can produce lots of new plants quickly and cheaply. But this is not the only reason that they do it. All the cuttings taken from one plant have identical DNA. They are genetically identical: we say that they are clones. So taking cuttings is a good way of reproducing a plant with exactly the features that you want. When flowering plants produce seeds, they are reproducing sexually. So new plants grown from seeds vary. They are not identical.

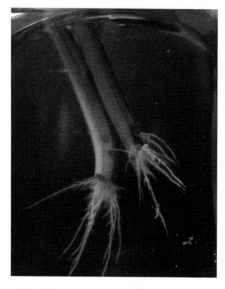

This willow cutting needed only water to grow.

Rooting powder

Questions

1 Name two parts of your body where you have stem cells.

2 Explain why a newt can regrow a leg but a human cannot.

3 For each of these types of cell, say whether they are fully unspecialized or not:
   a meristem cells
   b embryonic human stem cells
   c adult human stem cells

4 Give two reasons for growing plants from cuttings.

5 Explain how rooting powder helps a plant cutting to grow.

6 A gardener wants to grow dahlias with a variety of colours and sizes. Should she grow them from cuttings or seeds?
   Explain your answer.

Key words

unspecialized
stem cells
cuttings
rooting powder
auxins

Find out about:

▶ where genes are kept inside your cells

c A look inside the nucleus

All cells start their lives with a nucleus. A few specialized cells lose their nuclei when they finish growing. Human red blood cells are one example. Their job is to carry oxygen attached to haemoglobin molecules.

Red blood cells develop from stem cells in your bone marrow. As they develop, they make more and more haemoglobin. By the time they leave the bone marrow, they are full of haemoglobin and their nuclei have broken down.

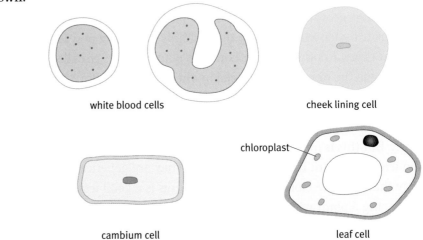

white blood cells　　　　　　　　　cheek lining cell

cambium cell　　　　　　　　　leaf cell

chloroplast

Cells vary in size and shape.

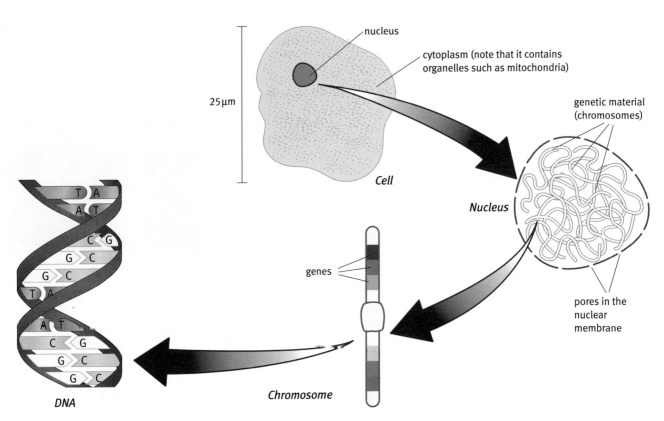

nucleus

cytoplasm (note that it contains organelles such as mitochondria)

25 μm

Cell

genetic material (chromosomes)

Nucleus

pores in the nuclear membrane

genes

Chromosome

DNA

Each chromosome in the nucleus contains thousands of genes.

Chromosomes

Imagine a chromosome as a long molecule of DNA wound around a protein framework. You have about a metre of DNA in each of your nuclei. This is made up of about 30 000 genes. Each gene probably codes for a protein.

Different species have different numbers of chromosomes and different numbers of genes (see the table on the right).

You have 23 pairs of chromosomes in your nuclei. You got one set of 23 from your mother's egg cell nucleus and the other set from the nucleus of your father's sperm.

Organism	Estimated gene number	Chromosome number
human	~30 000	46
mouse	~30 000	40
fruit fly	13 600	8
Arabidopsis thaliana (plant)	25 500	5
roundworm	19 100	6
yeast	6 300	16
Escherichia coli (bacterium)	3 200	1

These people are more alike than it appears. 99.9 % of their genes are the same.

Key words
chromosomes
genes

What is special about DNA?

You will find out later in the module that the molecule of DNA has a particular structure that allows it to:

▷ make exact copies of itself

▷ provide instructions so that the cell can make the right proteins at the right time

Questions

1 Name two ways in which your red blood cells are different from the other cells in your body.

2 Suggest why red blood cells wear out and have to be replaced every 2–3 months.

3 How many different types of protein are there likely to be in yeast?

4 Which animal has half as many genes as yeast?

5 To work as your genetic material, what two properties does DNA have?

Find out about:
▶ how your cells divide for growth and repair your body

D Making new cells

It is not always easy to tell whether something is a living organism or not.

You could ask yourself:

▶ Can it grow and reproduce?

▶ Is it made of cells?

Life cannot exist without the growth, repair, and reproduction of cells.

Cell division

When new body cells are made, they contain the same number of chromosomes as each other and the parent cell. They also contain the same cell parts, called **organelles**. So, before a cell divides, it must grow and make copies of:

▶ other organelles – such as **ribosomes** and **mitochondria**

▶ its nucleus, including the chromosomes

Only then does the cell divide. This part of the process is called **mitosis**.

Imagine a space probe bringing back objects like this from Mars. Scientists would need to find out whether they were alive or not. It might be a living organism able to colonize Earth!

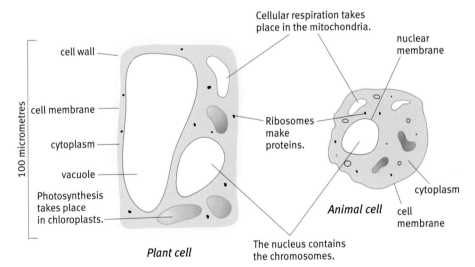

Cell organelles, such as mitochondria, are copied before a cell divides.

Copying chromosomes

You can see chromosomes by using a light microscope, but only in dividing cells. The DNA is too spread out in other cells. After the chromosomes are copied, the DNA strands become shorter and fatter. Then you can see them. You can read more about how DNA is copied in Section F.

single-stranded chromosome in non-dividing cell

double chromosome in dividing cell

You can see chromosomes only in dividing cells.

Mitosis

During mitosis, copies of chromosomes separate and the whole cell divides.

First, a complete set of chromosomes goes to each end of the dividing cell and forms two new nuclei. A complete set of organelles also goes to each end. Then the cytoplasm divides to form two identical cells. In plant cells, a new cell wall forms too.

1 Each chromosome has an identical copy attached to it. The membrane around the nucleus breaks down.

3 The chromosome copies separate to opposite ends of the cell.
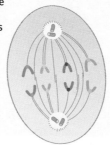

2 The chromosomes move to the centre of the cell.

4 New nuclear membranes form before the cell divides.
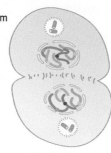

The pictures show what happens in mitosis in an animal cell. They show only 4 chromosomes.

Mitosis and asexual reproduction

Some plants and animals reproduce asexually. They use mitosis to produce cells for a new individual.

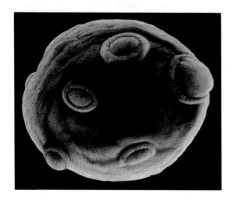

In yeast, new cells grow as buds from the parent.

Daffodil bulbs divide to form new ones.

This means that each of the individuals produced in asexual reproduction is genetically identical to the parent, so it is a clone of its parent.

Questions

1 What might scientists do to find out whether the objects from Mars in the drawing opposite are living things or not?

2 Before a cell divides, it grows. What two steps happen during cell growth?

3 What two main steps happen during cell division by mitosis?

4 a How many cells are made by mitosis?

b What are these new cells like compared to their parent cell?

Find out about:
▶ cell division to make gametes

E Sexual reproduction

Most plants and animals reproduce sexually. Males and females make sex cells or **gametes**, which join up at fertilization. The fertilized egg, or zygote, develops into the new life.

It is not always obvious which plant or animal is male and which is female. Some plants and animals are both.

It is easy to tell which is male and which is female.

The holly on the right is female. You cannot be sure about the one on the left.

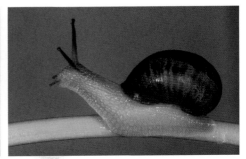

A snail has both male and female sex organs.

The only way to be sure about the sex of an organism is to look at its gametes. Males have small gametes that move. Females have large gametes that stay in one place.

Human males produce sperm in their testes. Females produce egg cells in their ovaries.

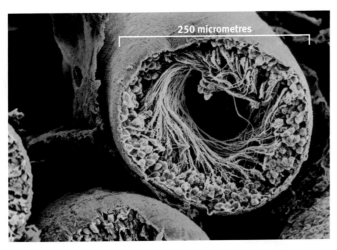

Sperm develop in tubules in the testes.

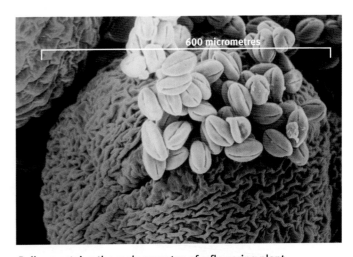

Pollen contains the male gametes of a flowering plant.

Male gametes are usually made in very large numbers. They move to the female gamete by swimming or being carried by the wind or an insect.

What is special about gametes?

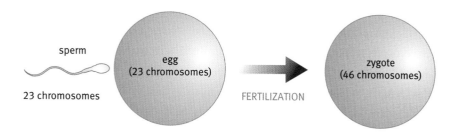

Meiosis halves the number of chromosomes in gametes. Fertilization restores the number in the zygote.

Human body cells have 23 pairs of chromosomes – 46 in total. Gametes have only 23 single chromosomes. This is important. When a sperm cell fertilizes an egg cell, their nuclei join up. So the fertilized egg cell (zygote) gets the correct number of chromosomes: 23 pairs – 46 in total. Gametes are made by a special kind of division called **meiosis**.

In humans, meiosis makes gametes that:

 have 23 single chromosomes (one from each pair)

 are all different – no two gametes have exactly the same genetic information

Offspring from sexual reproduction are different from each other and from their parents. We say that they show **genetic variation**.

Meiosis

Meiosis starts with normal body cells. It only happens in sex organs.

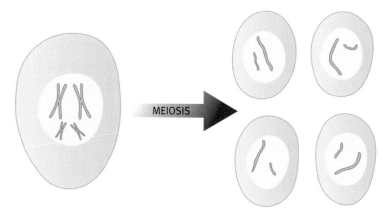

The chromosomes in this diagram have been copied. The parent cell divides twice, producing four cells. There are four cells after meiosis. They have half the number of chromosomes as the parent cell.

Key words
gametes
meiosis
genetic variation

Questions

1 Look again at the photos of holly at the top of the opposite page. You can be sure that the holly on the right is female. Why can you not be sure about the sex of the holly on the left?

2 Why are male gametes made in such large numbers?

3 Why is it important that gametes have only one set of chromosomes?

F The mystery of inheritance

In 1865 Gregor Mendel published his work on pea plants. You learnt about this in Module B3 *Life on Earth*. Mendel showed how features could be passed on from parents to their young. But there was still a mystery – how was the information passed from cell to cell? It took the work of many scientists to solve it.

1859	A chemical was extracted from nuclei and named 'nuclein'.
1944	'Nuclein' was recognized as genetic material.
Late 1940s	Erwin Chargaff discovered a pattern in the number of bases in DNA.
1951	Linus Pauling and Robert Corey showed that proteins have a helix structure.
1952–3	Rosalind Franklin and Maurice Wilkins produced X-ray diffraction pictures of DNA. They showed that the molecule had a regular, repeating structure.

Some of the discoveries that led up to the discovery of DNA structure.

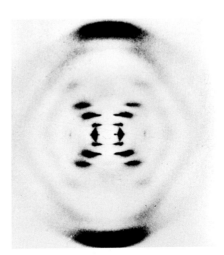

X-ray diffraction pictures of DNA, like this, show a repeating pattern.

Solving the mystery

In 1953, Francis Crick and James Watson published their now famous paper. 'A Structure for Deoxyribose Nucleic Acid', in the scientific journal *Nature* Their paper brought together all the work done on DNA. They used it work out the **double helix** structure of DNA.

Base pairing

There are four bases in DNA: adenine (A), thymine (T), guanine (G), and cytosine (C).

Erwin Chargaff had discovered that the amount of A is always the same as the amount of T, and the amount of G is the same as the amount of C. This is true no matter what organism the DNA comes from.

Crick and Watson concluded from this evidence that:

▶ A always pairs with T.

▶ G always pairs with C.

This is **base pairing**.

Watson (left) and Crick

The double helix

Watson made cardboard models of the bases. He found that A+T was the same size as G+C. Suddenly, he realized what that meant. In a molecular model, he and Crick fitted the pairs of bases between two chains of the other chemicals in DNA, a sugar called deoxyribose and phosphates. It worked. The shape turned out to be a double helix – a bit like a twisted ladder. Better still, it matched the X-ray evidence.

How DNA passes on information

Base pairing means that it is possible to make exact copies of DNA:

- Weak bonds between the bases split, opening up the DNA from one end to form two strands.

- Immediately, new strands start to form from free bases in the cell.

- As A always pairs with T, and G always pairs with C, the two new chains are identical to the original.

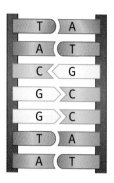

The rungs of the ladder are the pairs of bases held together by weak chemical bonds.

There are ten pairs of bases for each twist in the helix.

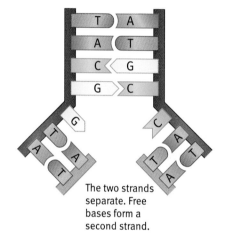

The two strands separate. Free bases form a second strand.

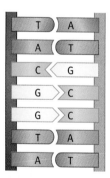

Each DNA molecule is made of half old DNA (black) and half new DNA (grey).

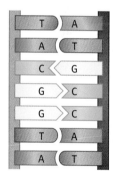

Each half of the split DNA molecule is complete, making two identical DNA molecules.

The structure of DNA (schematic)

Questions

1 The shape of a DNA molecule is sometimes described as a twisted ladder.

  a What makes the sides of the ladder?

  b What part of the ladder are the bases?

2 The bases in a DNA molecule always pair up the same way. Which base pairs with:

  a A?

  b C?

  c G?

  d T?

③ Describe what happens when a DNA molecule is copied.

4 Which observations made by other scientists did Crick and Watson's model account for?

Key words

double helix

base pairing

Find out about:

▶ some different proteins in your body

▶ why cells become specialized

G Specialized cells – special proteins

An oak tree has about 30 different types of cell. Your body has more than 300 types of cell. Each cell type has its own set of proteins.

Some proteins make up the framework of cells and tissues. These are **structural proteins**. If we take away all the water in an animal cell, 90% of the rest is proteins.

Protein	Found in . . .	Property
keratin	hair, nails, skin	strong and insoluble
elastin	skin	springy
collagen	skin, bone, tendons, ligaments	tough and not very stretchy

Different structural proteins have different properties.

The flesh of meat and fish is the animals' muscles. They are mainly protein and provide the protein in many people's diets. Plant seeds such as soya and other beans are rich in proteins and are the basis of many vegetarian meals.

Other proteins are essential for the chemical reactions that keep our bodies working. For example, **enzymes** speed up the chemical reactions in a cell. **Antibodies** are the proteins that help to defend us against disease.

Questions

1 Name three types of protein in your body.

2 Name one structural protein and say how it is suited to do its job.

An oak tree has about 30 cell types.

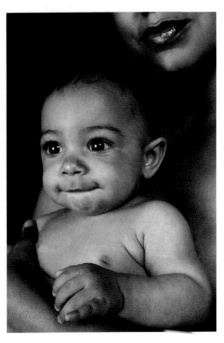

Humans have over 300 different types of cell.

What is the link between genes and proteins?

All the cells in your body come from just one original cell, the zygote. This divides to form a ball of cells. After about three weeks, cells start to specialize. They make the proteins needed to become a particular type of cell.

DNA is a cell's genetic code. Each gene is the instruction for a cell to make a different protein. By controlling what proteins a cell makes, genes control how a cell develops.

Each of your cells has a copy of all your genes. Something must make some cells turn into nerve cells and others into heart cells and all the other cell types. There must be **genetic switches**, but how they work is still a bit of a mystery.

Key words
structural proteins
enzymes
antibodies
genetic switches

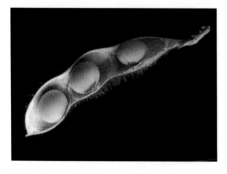

Rhino horn, tortoise shell, soya beans, steak . . . lots of different proteins.

Questions

3 DNA is a cell's genetic code. What does it do in a cell?

4 How do genes control how a cell develops?

5 Name a cell that would make these proteins:

  a collagen

  b amylase

  c haemoglobin

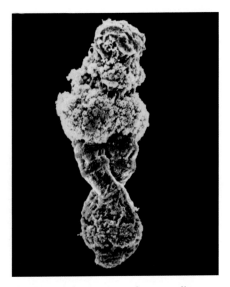

This is one chromosome from a salivary gland of a midge. The green areas are the active genes where DNA unravels whilst the protein is being made. One of these is instructing the cell to make a lot of amylase.

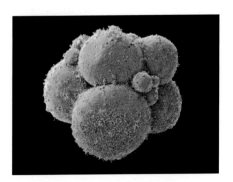

At the 8-cell stage, each cell of an embryo can develop into any kind of cell – or even into a whole organism.

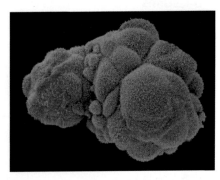

At six days and about 50 cells, a cell can develop into any kind of cell, but not into a whole organism.

Gene switching

The **one-gene-one-protein theory** says that each gene controls the manufacture of one type of protein. So, in any organism, there are as many genes as there are different types of protein. In humans there are 20 000 to 25 000 genes.

Not all these genes are active in every cell. As cells grow and specialize, some genes switch off.

In a hair cell, the genes for the enzymes that make keratin will be switched on:

hair cell genes switched on → enzymes for making keratin → hair grows

But the genes for those that make amylase will be switched off.

In a salivary gland cell:

salivary gland cell genes switched on → amylase secreted → starch digested

Gene switching in embryos

An early embryo is made entirely of embryonic stem cells. These cells are unspecialized. All the genes in these cells are switched on. As the embryo develops, cells specialize. Different genes switch off in different cells.

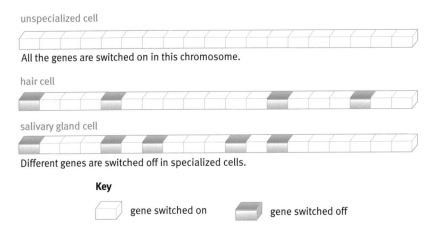

unspecialized cell

All the genes are switched on in this chromosome.

hair cell

salivary gland cell

Different genes are switched off in specialized cells.

Key

gene switched on   gene switched off

Some proteins are found in each type of cell, for example the enzymes needed for respiration. All cells respire, so the genes needed for respiration are switched on in all cells.

In adults, there are stem cells in parts of the body where there is regular replacement of worn out cells. These can develop only into cells of the particular organ where they are. So some of their genes must be switched off.

Right cell, right place

Compare the fingers on your right hand with the same fingers on your left hand. They are probably almost mirror images of each other. As we grow, the position and type of cells must be controlled, so each tissue and organ develops in the right place.

In some animals, such as frogs, the place where the sperm enters the egg affects where the head and tail of the animal will be. In mammals, the head and tail end are probably already decided. There are differences in the amount of proteins in different parts of the egg cell. This will affect which genes are active. In simple terms, genes for the front end will become active in one half of the embryo. Genes for the back end will be active in the other half. Early development of embryos is an important area of research.

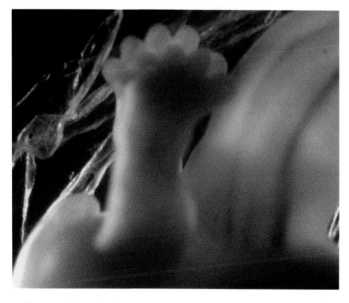

Cells near the end of a limb will make fingers. Cells nearer the body will make the arm. This happens because of the difference in the concentrations of chemical signals in each region of the embryo.

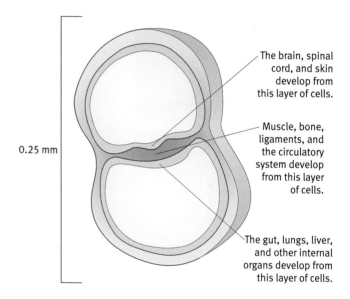

0.25 mm

The brain, spinal cord, and skin develop from this layer of cells.

Muscle, bone, ligaments, and the circulatory system develop from this layer of cells.

The gut, lungs, liver, and other internal organs develop from this layer of cells.

It is possible to map specialized parts of the body onto a diagram of a 14-day-old human embryo. Here you can see which groups of cells in the embryo will develop into future tissues and organs.

Questions

6 Suggest a function, other than respiration, that all cells carry out.

7 At the 8-cell stage of any embryo, how many genes are switched on?

8 What is the evidence that some genes are switched off at the 50–100-cell stage?

Key words

one-gene–one-protein theory

H Stem cells

A lot of research is going on into stem cells. This is because many scientists see the possibility of using them for:

▶ the treatment of some diseases

▶ the replacement of damaged tissue

Imagine if scientists could produce . . .	They might use them to treat . . .
nerve cells	Parkinson's disease and spinal cord injuries
heart muscle cells	damage caused by a heart attack
insulin-secreting cells	diabetes
skin cells	burns and ulcers
retina cells	some kinds of blindness

Scientists grew this skin from skin stem cells in sterile conditions. Doctors use it for skin grafts.

The problem is to find stem cells of the correct type and then to grow them to produce enough cells. Stem cells can come from early embryos, umbilical cord blood, and adults. Embryonic stem cells are the most useful because the cells are not yet specialized. All their genes are still switched on.

There are problems with this new technology. For example:

▶ Tissues from the embryonic stem cells do not have the same genes as the person getting the transpant. Transplanted tissue is rejected if your body recognizes that the cells are not from your body.

Cloning from your own cells

Another possibility is to remove the nucleus of a zygote and replace it with the nucleus from a patient's own body cell. The new embryo would have the same genes as the patient. So the embryonic stem cells produced would match those of the patient. There would be no problem of rejecting a transplant of tissues grown from these cells. This is **therapeutic cloning**. Some people do not agree that cloned embryos should be produced. You can learn more about this in Module B1 *You and your genes*.

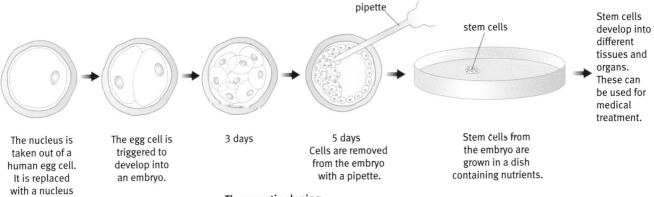

pipette

stem cells

Stem cells develop into different tissues and organs. These can be used for medical treatment.

The nucleus is taken out of a human egg cell. It is replaced with a nucleus

The egg cell is triggered to develop into an embryo.

3 days

5 days
Cells are removed from the embryo with a pipette.

Stem cells from the embryo are grown in a dish containing nutrients.

Therapeutic cloning

H Could adult stem cells be used instead?

Most adult human stem cells grow into only a few cell types because many genes are switched off. Suppose scientists could find a way to switch the genes back on. Then they could use a person's own stem cells to produce any cells they needed. But it is not easy to reactivate genes, and there are other problems with this technology. For example:

▶ Bone marrow, where blood cells form, is particularly rich in stem cells. But only 1 in 10 000 cells in bone marrow is a stem cell. So it is hard to separate them from the millions of other cells.

Some successes

Only very few people have been treated using stem cells. There have been some promising, but very early, results. Scientists from the University of Freiberg in Germany treated 60 patients who had heart disease. The patients were divided into two groups:

Group	Treatment	Improvement in the left ventricle after 6 months
1	injection of stem cells from the patient's own bone marrow into the heart muscle	7.0%
2	the best conventional treatment	0.7%

The scientists think that the stem cells

▶ turned into new blood vessel or heart muscle cells

▶ made the heart tissue secrete chemicals that encouraged growth of the patient's own heart cells

Questions

1 It may be possible to use a nucleus from a patient's own body cell to clone stem cells.

   a Describe how scientists hope to do this.

   b What is this type of cloning called?

   c What is the main problem with this technology?

2 What are the two main problems with using adult stem cells for therapeutic cloning?

3 Why is bone marrow rich in stem cells?

Key words

therapeutic cloning

113

Find out about:

▶ how DNA controls which protein a cell makes

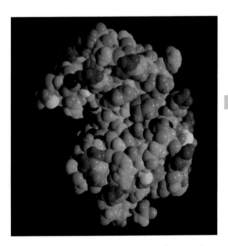

Enzymes work because of the shape of their active sites.

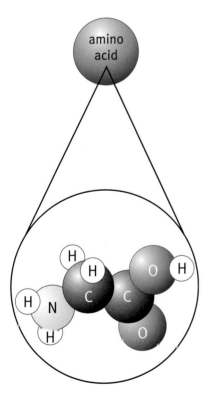

Amino acids are complex molecules. The diagrams on this page and the next represent the molecule simply.

Making proteins

The great number of jobs carried out by proteins means that they are very different from one another. The exact shape of a protein can be very important to how it works. Cells make proteins from about 20 different **amino acids**. They join them in chains of 50 to many thousands of amino acids. In each protein, the amino acids are joined in a particular order, but there are thousands of possibilities. The order of the amino acids fixes the way the chains of amino acids fold to form the three-dimensional shape of the protein.

The genetic code

When Crick and Watson discovered the structure of DNA, they were left with one major problem of the genetic code. They knew that DNA was the code for making proteins. But how could just four bases code for 20 amino acids?

If one base coded for one amino acid, DNA could code for only four amino acids. Two bases could code for 16.

In 1961, Crick worked out that a three-base code for each amino acid would work. This is called a **triplet code**. Different combinations of the four bases (A, T, G, and C) produce 64 triplet codes. So there is more than one code for each amino acid. There are also codes for start and stop. They mark the beginning and end of a gene.

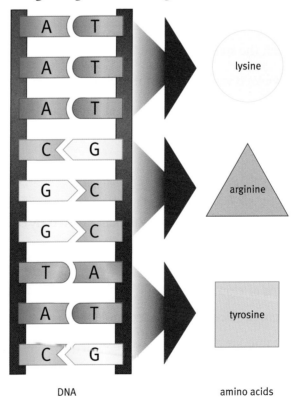

DNA   amino acids

Three bases on the DNA code for each amino acid. For example, TTT codes for lysine.

Which part of a cell makes proteins?

DNA in the nucleus contains the genetic code for making the proteins. But tiny organelles in the cytoplasm of a cell, called ribosomes, actually make them. Genes cannot leave the nucleus. So how do ribosomes get the instruction for making a protein? A molecule small enough to get through the pores of the nuclear membrane transfers the genetic code to the ribosomes. This smaller molecule is called messenger RNA (**mRNA**).

H The differences between DNA and mRNA are that mRNA has

- only one strand
- the base U in place of the T in DNA

The diagram below shows how a protein is made.

Key words

amino acids

triplet code

mRNA

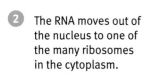

1 The gene unzips, and mRNA bases pair with DNA bases to form a strand of mRNA. U matches up with A instead of T.

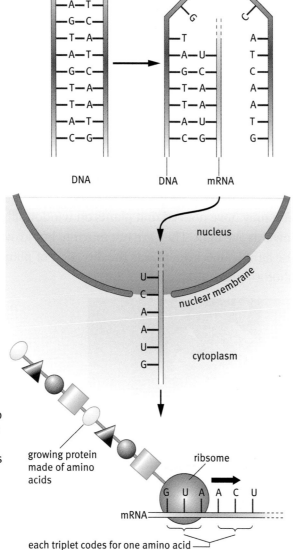

2 The RNA moves out of the nucleus to one of the many ribosomes in the cytoplasm.

3 The ribosome attaches to one end of the mRNA. As it moves along the mRNA, the ribosome reads the genetic code so that it can join the amino acids together in the correct order. When it has finished, the ribosome releases the protein into the cytoplasm and starts to make another one.

growing protein made of amino acids

ribosome

mRNA

each triplet codes for one amino acid

DNA makes protein with a ribosome's help (schematic).

Questions

1 Why are instructions for making proteins copied onto mRNA?

2 How many DNA bases code for an amino acid?

3 What is the triplet code for tyrosine?

4 Which amino acid has the code GCC?

5 Make bullet-point notes to explain how a protein is made. Your first bullet point should be:

- The gene unzips.

Your last bullet point should be:

- The ribosome releases the protein into the cytoplasm.

J Phototropism

Plants rooted in soil cannot move from place to place – not even the 'walking palm' tree in the picture below.

The walking palm tree, *Socratea durissima*, in Costa Rica. New roots grow towards a sunny patch and pull the stem and leaves towards the light. Hormones control the direction of root growth. Older roots in the shade die.

You may have noticed that plants on windowsills seem to be bending towards the light. They are not moving, but growing. When the direction the light comes from affects the direction of plant growth, it is called **phototropism**.

Questions

1 How does the walking palm tree grow towards the light?

2 Explain why a plant benefits from bending towards the light.

3 Write a definition for phototropism.

This houseplant has grown towards the window to increase the amount of light falling on its leaves.

H Darwin's phototropism experiments

Charles Darwin experimented with phototropism. He showed that the young shoots of grasses

- ❯ normally grew towards light

- ❯ remained straight when he covered their tips

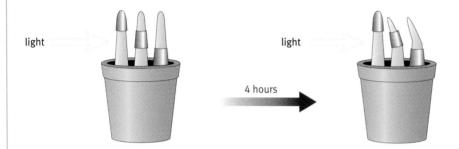

Foil covers different parts of the barley shoots.

In the experiment, shown in the picture above, covering the lower parts of the shoot did not stop bending towards the light. This shows that only the tip is sensitive to light. But the shoot bends below the tip – where cells are no longer dividing but are increasing in length.

Darwin did not know how bending towards light happened. Now scientists have found out that higher concentrations of auxins cause shoot cells to expand. The diagrams on the right explain how this causes phototropism.

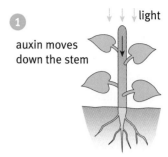

1 auxin moves down the stem

When a growing shoot gets light from above, the auxins spread out evenly and the shoot grows straight up.

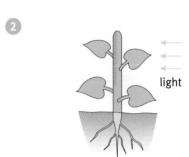

2

When light comes from one side, the auxins move over to the shaded side.

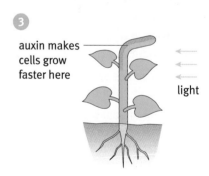

3 auxin makes cells grow faster here

The shoot grows faster on the shaded side, making the shoot bend towards the light.

How auxins explain phototropism

Questions

4 Look at the diagram showing phototropism experiments in barley. Suggest where:

   a the shoot detects light

   b the cells are growing very quickly to cause bending

5 a Which way will shoots A and B grow?

   b Will A or B grow taller? Explain your answer.

Key words

phototropism

117

B5 Growth and development

Summary

In this module you have met ideas about cell division and growth, and explanations for how a single fertilized egg cell can produce all the different cells needed for a new animal.

The cell cycle

▶ Cells go through cycles of growth and division:

- During cell growth the number of organelles increases, and chromosomes are copied.

- During cell division copies of the chromosomes separate, and then the cell divides.

▶ Cell division by mitosis produces two new cells identical to the parent cell.

▶ Cell division by meiosis leads to four gamete cells with half the chromosome number of the parent cell.

Making proteins

▶ Genes in the nucleus of a cell are instructions for making proteins.

▶ Copies of genes are carried from the nucleus to the cytoplasm, where proteins are produced.

▶ Chromosomes are made of the chemical DNA, which has a double helix structure that allows it to be reproduced accurately.

▶ DNA molecules are made up of four bases, which always pair up in the same way.

▶ The order of bases in a gene is the code for joining up amino acids in the correct order to make a particular protein. Ⓗ

A new human

▶ When two gametes join at fertilization, a zygote (fertilized egg cell) is produced that contains a set of chromosomes from each parent.

▶ Zygotes divide by mitosis to make an embryo.

▶ Up to the eight-cell stage of a human embryo, all the cells are identical embryonic stem cells.

▶ Embryonic stem cells can produce any sort of cell required by the organism.

▶ Cells become specialized because many genes are not active, so the cell only produces the particular proteins that it needs.

▶ It is possible to reactivate some inactive genes in the nucleus of body cells. Ⓗ

▶ Adult and embryonic stem cells have the potential to produce cells needed to replace damaged tissues. Ⓗ

Plant development

▶ Plant cells specialize into plant tissues (e.g. xylem, phloem) and organs (e.g. roots, leaves, flowers).

▶ Some plant cells (meristem cells) stay unspecialized throughout the plant's life, so they can develop into any type of plant cell.

▶ Mersitem cells divide to make new cells that increase the plant's height, width, and length of roots.

▶ Meristem cells can re-grow whole parts of the plant, which means that gardeners can make new plants from cuttings.

▶ Plant hormones (e.g. auxins) affect how plant cells develop.

▶ Plants grow towards light; this response is called phototropism.

▶ Auxin builds up in cells on a shoot's shady side, which makes the cells grow longer, so the shoot bends towards the light. Ⓗ

Questions

1 This question is about the cell cycle of a human body cell.

  a Write down what happens during:

    i growth phase

    ii cell division

  b Copy and complete the table below, to describe mitosis in a human body cell.

Why mitosis happens	growth
Where mitosis happens	
Number of cells after division	
Number of chromosomes in new cells	

  c Draw a similar table to describe meiosis.

2 **a** Describe the shape of a DNA molecule.

  b Draw a flow chart to explain how a gene controls the production of a protein. The first and last statements are:

    • Genes are the code for producing a protein. They are kept in the nucleus.

    • The new protein is made.

3 **a** Name the four bases that make up a DNA molecule.

  b Describe how bases pair up in a DNA molecule.

  c Draw a flow chart to explain how a DNA molecule is copied. The first and last statements are:

    • A DNA molecule is made of two strands.

    • Two new DNA molecules have been made.

4 **a** Name two tissues from:

    i an animal

    ii a plant

  b Name the unspecialized cells in:

    i an animal

    ii a plant

  c Explain why many plants can re-grow whole organs if they are damaged, but most animals cannot.

5 Genes are instructions for making proteins. All human body cells contain the same genes. But cells only make the proteins that they need to do their specialized job.

Explain how cells with the same genes can become specialized.

6 **a** Name the plant hormone found in rooting powders.

  b Explain why young plants grown on a window sill grow towards the light.

Why study chemicals of the natural environment?

Conditions on Earth are special. The temperature is just right for most water to be liquid. The atmosphere has enough oxygen for living things to breathe but not so much that everything catches fire. Rocks are the source of many of the chemicals that meet our daily needs.

The science

Theories of structure and bonding explain how atoms are arranged and held together in all the chemicals and materials that make up our Earth. There are three types of strong bonding that give rise to the useful materials that are metals, polymers, and ceramics. There are weaker forces of attraction that allow molecules to stick together enough to make liquids and solids.

Chemistry in action

Modern life depends on a wide range of advanced materials that have been developed as scientists understand more about structure and bonding. The science of structure and bonding is now so advanced that scientists can do engineering on an atomic scale, called nanotechnology.

Scientists are applying their understanding of chemistry in the environment to work out how society can deal with the growing impacts of human development.

Chemicals of the natural environment

Find out about:

- the chemicals in the main spheres of the Earth
- theories of structure and bonding
- the impact of human activity on the chemistry of the environment
- methods used to extract metals from their ores
- how to calculate chemical quantities

Find out about:

- naturally occurring elements and compounds
- abundances of elements in different spheres
- cycling of elements between the spheres

A Chemicals in four spheres

People who study the Earth often think of it as being made up of spheres (see the diagram below). Starting from the middle, first comes the core, then the mantle, then the crust.

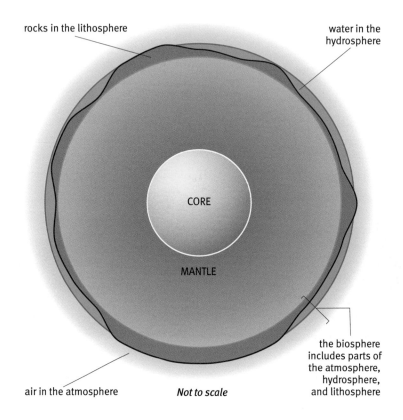

rocks in the lithosphere

water in the hydrosphere

CORE

MANTLE

air in the atmosphere

Not to scale

the biosphere includes parts of the atmosphere, hydrosphere, and lithosphere

The Earth's spheres. The spheres are much smaller than this relative to the core and mantle.

The **lithosphere** is broken into giant plates that fit around the globe like puzzle pieces. These are the tectonic plates which move a little bit each year as they slide on top of the upper part of the mantle. The **crust** and upper **mantle** make up the lithosphere. They are about 80 km deep.

The oceans and rivers make up the **hydrosphere**. This is not a complete sphere, but the oceans cover two-thirds of the globe, so it almost is.

There are living things in the sea, rivers, and lakes. Large areas of land are covered with living things, and there are billions of living things in soil. So scientists like to think of a sphere of life encasing the planet; they call this the **biosphere**.

Finally, wrapped like a big fluffy duvet around the Earth, keeping it warm, is the layer of air we call the **atmosphere**.

Key words

lithosphere
crust
mantle
hydrosphere
biosphere
atmosphere

Elements in the spheres

The spheres vary greatly in their chemical composition. The air consists mainly of two free elements: nitrogen and oxygen. The hydrosphere is almost entirely the compound water.

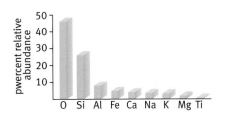

The abundant elements in the lithosphere

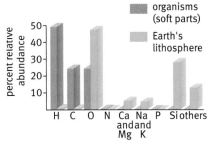

The percentage of different elements in living tissue and in the lithosphere compared

The rocks of the lithosphere are made up mainly of silicates. These are compounds of silicon and oxygen together with much smaller quantities of other elements.

The elements exist as different chemical species in the four spheres. In the lithosphere, carbon combines with hydrogen to make the hydrocarbons in crude oil. Carbon is also hidden in the calcium carbonate of chalk and limestone. In the atmosphere, carbon is present as the gas carbon dioxide. Carbon is a vitally important element in the biosphere: most of the chemicals that make up living things are compounds of carbon with hydrogen, oxygen, and a few other elements.

Flowing between the spheres

Chemicals do not always stay in one sphere. They are constantly on the move between the spheres (see the diagram on the right). Think of a carbon atom: it may start in the atmosphere; be taken into a plant in the biosphere; be washed into water in the hydrosphere; then buried in sediment of the lithosphere.

Water flows freely between the spheres. The obvious place for water is the hydrosphere. But think of clouds in the atmosphere; then rain sinking into the lithosphere, reappearing as a spring, where it gets drunk by an animal or soaked up by roots back into the biosphere.

Questions

1. 'The biosphere overlaps with the atmosphere, hydrosphere, and lithosphere.' Give some examples to illustrate this statement.

2. Make a list of 10 different things you use (including what you wear and eat) during a day. Identify which sphere each item has come from.

3. Look at the bar charts on this page. What are the three most abundant elements in living tissue? What are the three most abundant elements in the lithosphere? Comment on your answers.

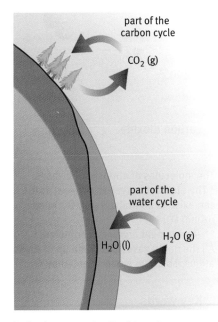

Cycles between the Earth's spheres. The most active places are the junctions between the spheres. Photosynthesis happens at the junction between the atmosphere and biosphere. Volcanoes move gases from the lithosphere to the atmosphere and the hydrosphere.

Find out about:

▸ gases in the air
▸ weak attractions between molecules
▸ strong covalent bonding

B Chemicals of the atmosphere

The Earth is just the right size for its gravity to hold on to gases. It is also just the right distance from the Sun to have the right temperature for liquid water to exist. This water, together with the carbon dioxide and oxygen in the atmosphere, means that Earth can support a great variety of plant and animal life.

The composition of the atmosphere is 78% N_2, 21% O_2, 1% Ar, 0.03% CO_2, with small amounts of water vapour.

The presence of argon in the air was not discovered until the 1890s – over a hundred years after the discovery of oxygen. Argon is a noble gas but it is not a rare gas. Every time you breathe in you take about 5 cm^3 argon into your lungs.

All the chemicals in the atmosphere are gases at normal temperatures – that is why they have ended up in the atmosphere. This means they have low melting and boiling points.

The other similarity between the chemicals in the atmosphere is that they are all either non-metallic elements (O_2, N_2, Ar), or they are compounds made from atoms of non-metallic elements. For example, CO_2 is made from carbon and oxygen.

Atoms and molecules in the air

Most of the chemicals in the atmosphere are made of **small molecules**. Only the noble gases exist as single atoms.

All molecules have a slight tendency to stick together. For example, there is an attraction between one O_2 and another O_2. But these **attractive forces** are weak. This is why the chemicals that make up the atmosphere are gases with low melting and boiling points.

One way to picture this is that the molecules are moving so quickly that, when two O_2 molecules come close to each other, the attractive force between them is not strong enough to hold them together.

Strong bonds in molecules

The forces inside molecules that hold the atoms together are very strong, many times stronger than the weak attractions between molecules. Small molecules such as O_2 or H_2 do not split up into atoms except at very, very high temperatures.

nitrogen	
oxygen*	
argon**	
carbon dioxide	

The molecules of atmospheric gases.
* The element oxygen is unusual as it can exist as normal oxygen, O_2, or as ozone, O_3.
** The element argon is a noble gas. All the noble gases are made up of single atoms, so argon is represented by Ar.

Key words

small molecules
attractive forces
molecular models
electrostatic attraction
covalent bonding

Chemists often use a single line to represent a single bond between atoms in a molecule. For example, the simple molecules in hydrogen, oxygen, and carbon dioxide can be represented by the molecular formulae H_2, O_2, and CO_2. But if you want to show their bonds, they can be represented by:

$$H - H \qquad O = O \qquad O = C = O$$

Some **molecular models** use the same idea. A coloured ball represents each atom, and a stick or a spring is used for each bond joining them together.

Electrons and bonding

A knowledge of atomic structure can help to explain the bonding in molecules (see Module C4 *Chemical patterns*, Section K). When non-metal atoms combine to form molecules, they do so by sharing electrons in their outer shells.

The atoms are held strongly together by the attraction of their nuclei for the pair of electrons they share.

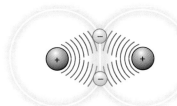

The atoms cannot move any closer together because the repulsion between the two positively charged nuclei will push them back apart again.

The formation of a covalent bond between two hydrogen atoms

The molecule of hydrogen is held together by the **electrostatic attraction** between the two nuclei and the shared pair of electrons. This is a single bond.

This type of strong bonding is called **covalent bonding**. 'Co' means 'together' or 'joint', while the Latin word 'valentia' means strength. So we have strength by sharing.

The number of bonds that an element can form depends on the electrons in the outer shell of its atoms. The table on the right gives the number of covalent bonds normally formed by atoms of some common non-metal elements.

H₂

O₂

H₂O

CO₂

Ball-and-stick models of H_2, O_2, H_2O and CO_2. Notice that the molecules have a definite shape. There are fixed angles between the bonds in molecules

---

Questions

1 Estimate the volume of argon in the room in which you are sitting.

2 Draw diagrams to show the covalent bonding in these molecules:

    a hydrogen chloride, HCl
    b ammonia, NH_3
    c methane, CH_4
    d ethene C_2H_4

Atom	Usual number of covalent bonds
H, hydrogen	1
Cl, chlorine	1
O, oxygen	2
N, nitrogen	3
C, carbon	4
Si, silicon	4

Find out about:

▶ unusual properties of water
▶ bonding within and between water molecules
▶ ions in solution

On Earth, water exists mostly in the liquid state (the oceans and the clouds), with some in the solid state (ice) and a smaller amount as gas (water vapour).

Key words

salts
dissolve
weathering

c Chemicals of the hydrosphere

Properties of water

Water is such a familiar chemical that it not obvious that chemically it has some very special properties.

One of these special properties is that it is a liquid at room temperature. The H_2O molecule has a smaller mass than molecules of O_2, N_2, or CO_2, which are all gases at normal temperatures. So you might expect H_2O to be a gas at normal temperatures. But H_2O melts at 0 °C and boils at 100 °C. For some reason, water molecules of H_2O have a greater tendency to stick together than the molecules of a gas like oxygen.

Strange things happen when water cools from room temperature. At first, the liquid contracts as expected. This goes on until the temperature falls to 4 °C. Then, on cooling further to 0 °C, it starts to expand. This means that ice is less dense than the cold water surrounding it, so it floats. This is very important in nature. It means that winter ice does not sink to the bottom of lakes where it would be far from the warming rays of spring sunshine.

A third odd property of water is that it is a good solvent for **salts**. Most common solvents do not **dissolve** ions, but water does.

Pure water does not conduct electricity – in this respect it resembles other liquids that are made up of small molecules. This shows that it does not contain charged particles that are free to move. Even though pure water does not conduct, you should never touch electrical devices with wet hands. This is because the water on your hands is not pure, so it does conduct electricity and increases the risk of you receiving an electric shock.

Water molecules

Knowledge of the structure of water molecules can help to explain its remarkable properties. The three atoms in a water molecule are not arranged in a straight line but are at an angle.

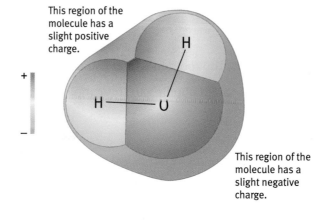

This region of the molecule has a slight positive charge.

This region of the molecule has a slight negative charge.

A water molecule, showing that oxygen has a slightly greater share of the pair of electrons in each bond

In the covalent bonds between the atoms, the electrons are not evenly shared. The oxygen atoms have more than their fair share. This means that there is a slight negative charge on the oxygen side of each molecule and a slight positive charge on the hydrogen side. Overall the molecules are still electrically neutral.

The small charges on opposite sides of the molecules cause slightly stronger attractive forces between the molecules. These small charges also help water dissolve ionic compounds by attracting the ions out of their crystals.

The attractions between water molecules and their angular shape mean that in ice they line up to create a very open structure. As a result, ice is less dense than liquid water.

Why is sea water salty?

The diagram below shows how soluble chemicals get carried from rocks to the sea during part of the water cycle.

The main soluble chemical carried into the sea is sodium chloride. River water does not taste salty because the concentration is so low, but the concentration of salt in the sea has built up over millions of years, and so it tastes salty.

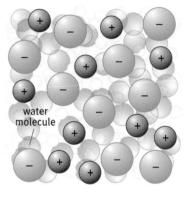

water molecule

Ions separate and can move about freely when dissolved in water. In sea water there is a mixture of positive ions and negative ions.

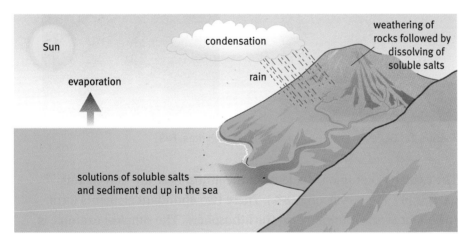

Weathering slowly breaks down rocks. This exposes the inside of the rocks to water. Soluble chemicals in the rocks dissolve in the water and get washed away.

Other chemicals that dissolve include potassium chloride, potassium bromide, potassium iodide, magnesium chloride, and magnesium sulfate. One litre of typical sea water contains about 40 g of dissolved chemicals from rocks.

Most of the compounds that are dissolved in sea water are made up of positively charged metal ions and negatively charged non-metal ions. They are salts.

Questions

1. How would the possibility of life on Earth have been affected had ice been denser than water?

2. Draw up a table to list in one column the properties of water that are typical of molecular chemicals and the properties that are unusual.

3. Write out the formulae of these salts with the help of the table showing the ion charges on page 59 in Module C4. The sulfate ion is $SO_4{}^{2-}$.
   a potassium iodide
   b magnesium chloride
   c magnesium sulfate

Find out about:

▶ chemicals in the crust of the Earth
▶ ionic bonding
▶ silica and silicates
▶ giant covalent structures

D Chemicals of the lithosphere

Rocks and minerals

The outer rigid layer of the Earth is the lithosphere. 'Lithos' is the Greek word for stone or rock. The top part of the lithosphere – the part we live on – is the crust.

Some rocks are massive, like the cliffs on the side of a mountain. Scientists include boulders, stones, and pebbles as rocks. A stone is just a small piece of rock.

Rocks are made of one or more minerals. Sandstone, for example, is made of mainly one mineral: silicon dioxide. Limestone is also made of a single mineral: calcium carbonate. Granite is a mixture of quartz, feldspar, and mica.

Minerals are naturally occurring chemicals. They may be elements, like gold and silver, which are found free in rocks. More commonly, they are compounds, such as silicon dioxide, SiO_2; calcium carbonate, $CaCO_3$; rock salt, $NaCl$; and iron oxides such as Fe_2O_3.

Granite is made from a mixture of minerals. There are glassy grains of silica (silicon dioxide), black crystals of mica, and large crystals of feldspar, which may be pink or white.

Haematite, Fe_3O_4. Crystals of this mineral range from metallic black to dull red.

Calcite, $CaCO_3$

Pyrites, FeS_2

The two most common elements in the lithosphere are non-metals: oxygen (47%) and silicon (28%). These two **abundant** elements form the major types of minerals in the lithosphere. The simplest example is silicon dioxide, SiO_2, which can take various crystalline forms, including quartz.

Key words

granite
mineral
abundant

Questions

1 Write a sentence that makes clear the difference between the words *rock*, *mineral*, and *lithosphere*.

2 Look at the graphs on page 123.
   a What is the most abundant metal in the lithosphere?
   b What is the most abundant non-metal?

3 Name a mineral which is:
   a an oxide   b a chloride   c a sulfide
   d a carbonate

4 Name the elements combined together in:
   a quartz   b galena   c calcite
   d pyrite

Evaporite minerals

Sea water contains abundant dissolved chemicals. When the water evaporates, **ionic compounds** crystallize. Sodium chloride, NaCl, or rock salt, is a common example. The mineral is halite.

Minerals that are formed in this way are called evaporites. Roughly 200 million years ago, in the Triassic era, vast salt deposits were laid down as sea water evaporated. The salt was later covered with other sediments and is now under the county of Cheshire in England. People have been extracting and trading this salt since before Roman times.

The structure and properties of salts

Crystals of sodium chloride are cubes. They are made up of sodium and chloride ions (see Module C4 *Chemical patterns*, Section J).

In every crystal of sodium chloride, the ions are arranged in the same regular pattern. This means that all crystals of the compound have the same cubic shape.

As a sodium chloride crystal forms, millions of Na$^+$ ions and millions of Cl$^-$ ions pack closely packed together. The ions are held together very strongly by the attraction between their opposite charges. This is called **ionic bonding**, and the structure is called a **giant ionic structure**. Unlike compounds such as water which are made up of individual molecules of H$_2$O, there is *not* an individual NaCl molecule.

Because of the very strong attractive forces, it takes a lot of energy to break down the regular arrangement of ions. So NaCl has to be heated to 808 °C before it melts. Compare this to melting ice (melting point 0 °C), where you only need to supply sufficient energy to overcome the relatively weak attractive forces between the H$_2$O molecules.

Salt deposits in the Great Basin. In this hot and dry area of the western USA, the salty water evaporates so fast that the salts crystallize as salt flats.

1 The Na$^+$ also attracts other Cl$^-$ ions that are close to it. The Cl$^-$ attracts other Na$^+$ ions that are close to it.

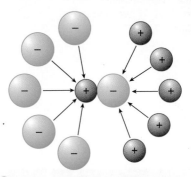

2 Another 5 Cl$^-$ ions can fit around the Na$^+$, making 6 in total, and another 5 Na$^+$ ions can fit around the Cl$^-$ ion, making 6 in total.

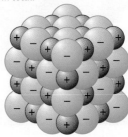

3 Each of these ions then attracts other ions of the opposite charge, and the process continues until millions and millions of oppositely charged ions are all packed closely together.

How a sodium chloride crystal forms.

Questions

5 Draw up a table to compare the properties of a molecular compound such as water and an ionic compound such as sodium chloride.

Key words

ionic compound
ionic bonding
giant ionic structure

Silica and silicates

Quartz

The mineral silica consists of silicon dioxide, SiO_2. Its commonest crystalline form is **quartz**. Crystalline silica has helped to shape human history. From the sand used for making glass, to the piezoelectric quartz crystals used in advanced communication systems, crystalline silica has been a part of human technological development.

Pure quartz crystals. Pure SiO_2 is transparent and very hard.

SiO_2 is found in many rocks. As the rocks get weathered, the SiO_2 stays behind and ends up as sand in rivers and beaches. When sand is compressed, it forms the rock called **sandstone**.

When silicon atoms bond to other atoms, they normally form four bonds per Si atom. Oxygen, on the other hand, normally forms two bonds per O atom. So, in SiO_2, each Si atom forms a covalent bond to four O atoms. Each O atom forms a covalent bond to two Si atoms.

The diagram below shows the arrangement of the Si and O atoms when they have bonded together in SiO_2. You can see that, instead of forming small molecules, they form a three-dimensional **giant covalent structure** that goes on and on. The Si—O covalent bond is very strong, so this giant structure is very strong and rigid. This is what gives SiO_2 its special properties (see the table on page 131).

Amethyst is a form of quartz used as a **gemstone** in jewellery. It contains traces of manganese and iron oxides, which give the mineral its violet colour.

Key words

quartz
sandstone
gemstone
giant covalent structure

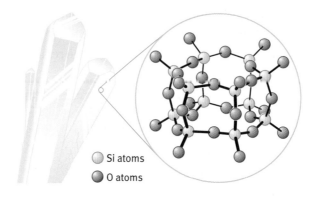

○ Si atoms
● O atoms

In SiO_2, each Si atom (grey) is covalently bonded to four O atoms (red). Each O atom is bonded to two Si atoms. On average there are four O atoms to every two Si atoms, so the formula simplifies to SiO_2.

Property of SiO$_2$	Comments	Uses
very hard	strong rigid structure; will scratch steel	used as abrasive in sandpaper and scouring powders
high melting (1610 °C) and boiling (2230 °C) points	strong, rigid structure difficult to break down	used to make linings for furnaces and high-temperature laboratory glassware
insoluble in water	resists weathering, ending up as sand in rivers, on beaches, and in deserts	sandstone used as building stone
electrical insulator	no free electrons or ions to carry electricity	silica glass used as an insulator in electrical devices

Some properties of SiO$_2$

Questions

6 What properties of silicon dioxide make it useful in sandpaper?

7 What properties of quartz as amethyst make it suitable for a gemstone?

8 Show how the properties of silicon dioxide can be explained in terms of its structure and bonding.

Silicate minerals

Over 95% of the rocks in the Earth's continental crust are formed by silica and the silicate minerals. However, the other 5% includes some very important minerals such as limestone and gold.

The Deccan Traps in India show how igneous rocks made of silicates can shape the landscape on a huge scale. The Traps cover an area of 500 000 square kilometres. They rise to over 2000 metres above sea level and consist of layer upon layer of basalt. These are the remains of lava flows which spread out about 66 million years ago. Basalt is an igneous rock which consists of tiny grains of a mixture of silicate minerals.

The structure of silicate minerals is based on giant structures of silicon and oxygen atoms. They include atoms of other elements too such as aluminium, iron, calcium, and potassium. The mica and feldspar in granite are silicate minerals.

Find out about:

▶ chemicals in living things
▶ proteins, carbohydrates, and DNA

E Chemicals of the biosphere

All living things are made from chemicals. The study of the chemicals of life is called biochemistry. Most biochemicals are based on three elements: carbon, hydrogen, and oxygen, sometimes with nitrogen, sulfur and phosphorus. The compounds consist of large molecules.

Why carbon?

The skeleton of all biochemicals is made from carbon. Carbon is the element on which all life is based. The special things about carbon that make this possible are:

▶ Carbon atoms can form chains by joining to themselves.

▶ Carbon forms four strong covalent bonds, so other atoms can join onto the chains (see below). Very often these are hydrogen atoms.

These properties of carbon mean it can make an amazing variety of compounds. This is why life itself is amazingly varied. Most biochemicals are polymers, built by joining together smaller, simpler molecules.

glycine

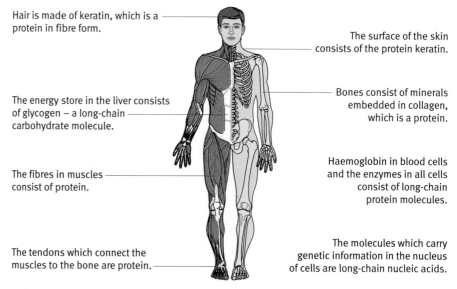

Hair is made of keratin, which is a protein in fibre form.

The surface of the skin consists of the protein keratin.

The energy store in the liver consists of glycogen – a long-chain carbohydrate molecule.

Bones consist of minerals embedded in collagen, which is a protein.

Haemoglobin in blood cells and the enzymes in all cells consist of long-chain protein molecules.

The fibres in muscles consist of protein.

The tendons which connect the muscles to the bone are protein.

The molecules which carry genetic information in the nucleus of cells are long-chain nucleic acids.

Polymers in the human body

Proteins

Hair, skin, and muscle are built from **proteins**. The enzymes that control biochemical reactions are made of proteins. Everywhere you look in your body, you will find proteins doing different kinds of jobs. Proteins are so varied because the carbon chains they are built from can be so varied.

Proteins are polymers, built by joining together monomers called amino acids. There are 20 different amino acids, and proteins have hundreds of them joined together.

alanine

Models of two amino acid molecules and their structures shown in symbols

Each protein has a unique sequence of amino acids in its polymer chains. This is shown in Module B5 *Growth and development*, page 115.

Carbohydrates

Photosynthesis in the leaves of plants turns carbon dioxide and water into glucose. Glucose is a sugar, which belongs to the family of compounds called **carbohydrates**.

There are just three elements in carbohydrates. They are compounds of carbon and the elements in water. This explains their name:

'carbo-'   for carbon

'-hydrate'   from the Greek word for water

Glucose is a very soluble sugar. Plants convert it to starch as an energy store. Starch is an insoluble polymer made up of long chains of glucose units.

Plants also join up glucose molecules in a different way to produce another polymer called cellulose. This is the polymer that makes up plant cell walls.

Nucleic acids

DNA and RNA are nucleic acids. They are the molecules that carry the genetic code.

The backbone of a DNA molecule is a polymer with alternating sugar and phosphate groups. Attached to the backbone are four bases: adenine, cytosine, guanine, and thymine.

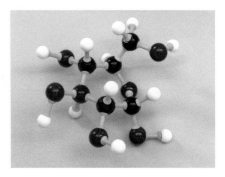

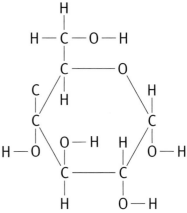

A model of a glucose molecule and its structure in symbols

A model of a short length of a DNA molecule. The blue atoms represent nitrogen and the violet atoms represent phosphorus. A schematic way of representing a DNA molecule is shown in Module B5, on page 107.

Questions

1 Look at the models of three amino acid molecules.

   a Which five elements are present in the molecules?

   b In these molecules, how many bonds are formed by:

   i each carbon atom?
   ii each oxygen atom?
   iii each nitrogen atom?

   c Write the molecular formulae for the three amino acids

   d In what ways are the structures of the amino acids the same, and how do they differ?

2 Name the three elements which make up carbohydrates.

3 If the formula of glucose is written $C_xH_yO_z$, what are the values of x, y, and z?

4 Look at the model of a DNA molecule on the right. Which elements are present in DNA?

Key words

protein
photosynthesis
carbohydrate
DNA

133

Find out about:
- natural cycles of elements
- human impacts on the environment

Questions

1 Give an example of a chemical species containing carbon in:
   a the atmosphere
   b the hydrosphere
   c the lithosphere
   d the biosphere

2 According to the figures in the diagram of the carbon cycle, what is the total mass, in gigatonnes, of carbon in the biosphere?

3 **a** What is the total mass of carbon passing into the atmosphere each year?
   b What is the total mass of carbon taken out of the atmosphere each year?
   c What is the net change in the mass of carbon in the atmosphere each year?

4 **a** What mass of carbon is removed from the atmosphere each year by the photosynthesis of plant life on land?
   b What mass of carbon is added to the atmosphere each year by the respiration of living things on or in the land?

5 Does the diagram of the carbon cycle suggest that human activity is having a significant effect on the global carbon cycle?

F   Human impacts on the environment

Element cycles

As living things grow, die, and decay, the elements move between the biosphere, hydrosphere, atmosphere, and lithosphere. This happens naturally on a large scale.

The carbon cycle

Some human activities can make a difference to these **natural cycles**. An important example is the effect on the **carbon cycle** of burning fossil fuels.

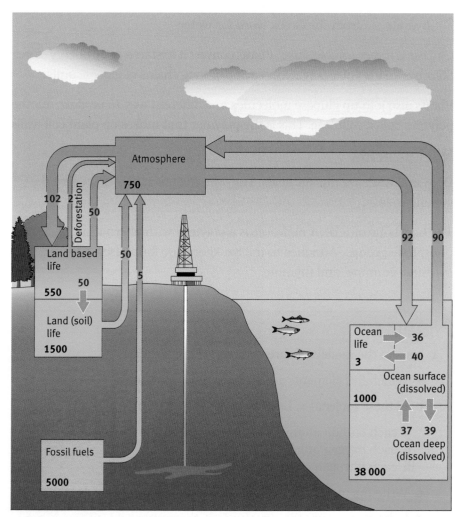

A simplified version of the carbon cycle. The figures are in 1000 million tonnes (gigatonnes). The figures in black are estimates of the total mass of carbon (worldwide) in the different spheres. The figures in red are the estimated flows of carbon between spheres.

The flow of carbon dioxide into the air from burning fuels is small compared with the natural flows due to respiration and photosynthesis. Even so, it is large enough to have raised the concentration of carbon dioxide in the air from about 277 parts per million before the Industrial Revolution to around 360 parts per million today.

The nitrogen cycle

Nitrogen is essential for making many biochemicals, especially proteins, which are the main nitrogen-containing compounds in the biosphere.

N_2 (nitrogen gas) has small molecules with weak bonds between them. It is a gas and is found in the atmosphere.

NO_3^- (nitrate) and NH_4^+ (ammonium) are ions. They are attracted to water molecules, which makes them dissolve. So they may get into the hydrosphere. They are also attracted to ions in the soil, so they get into the lithosphere. When they are absorbed by plant roots, they enter the biosphere.

Growing crops take in the nitrogen they need from the soil in the form of nitrate ions. Farmers and gardeners add fertilizers to the soil. This puts back the nitrogen removed from the soil when crops are grown and harvested.

The problem for intensive agriculture is that plants cannot use nitrogen directly from the air. Only a few specialized bacteria and algae can convert N_2 gas to ammonium ions and nitrates. This is the process called 'fixing' nitrogen. Nitrogen fixation makes the nitrogen available for use by plants.

There are three main ways of fixing nitrogen from the air:

- the action of microorganisms (bacteria or algae)

- a chemical reaction in the air during lightning flashes

- a manufacturing process called the Haber process used to make fertilizers

The manufacture of fertilizers now makes a major impact on the **nitrogen cycle**. As much nitrogen is fixed by industry as is fixed naturally by the natural processes supplying nitrogen to the soil.

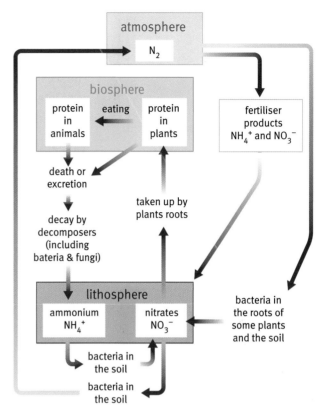

Nitrogen gets converted to different chemical species as it goes round the nitrogen cycle.

Key words

natural cycle
carbon cycle
nitrogen cycle

Questions

6 Explain why:
   a there is lots of N_2 in the atmosphere, but very little in the lithosphere
   b there is lots of NH_4^+ in the lithosphere, but very little in the atmosphere

7 Explain why plants can be short of nitrogen when there is so much nitrogen in the air.

8 Suggest possible consequences of the large-scale fixing of nitrogen by industry and the use of synthetic nitrogen compounds as fertilizers.

G Metals from the lithosphere

Metal ores

The wealth of societies has often depended on their ability to extract and use metals. Mining and quarrying for metal ores takes place on a large scale and can have a major impact on the environment.

All metals come from the lithosphere, but most metals are too reactive to exist on their own in the ground. Instead, they exist combined with other elements as compounds. Like other compounds found in the lithosphere, they are called minerals

Rocks which contain useful minerals are called **ores**. The valuable minerals are very often the **oxides** or sulfides of metals.

Metal	Name of the ore	Mineral in the ore
aluminium	bauxite	aluminium oxide, Al_2O_3
copper	copper pyrites	copper iron sulfide, $CuFeS_2$
gold	gold	gold, Au
iron	haematite	iron oxide, Fe_2O_3
sodium	rock salt	sodium chloride, NaCl

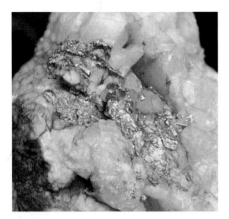

Gold is so unreactive that it occurs uncombined in the lithosphere. But most metals occur as compounds.

An open-pit copper mine in Utah, USA. Mining on this scale makes a big impact on the environment.

Because it occurs in an uncombined state, gold has been used by humans for more than 5000 years. More **reactive metals** like iron were not used by humans until methods for extracting them had been developed.

Mineral processing

Over hundreds of millions of years, rich deposits of ores have built up in certain parts of the Earth's crust. But even the richest deposits do not contain pure mineral. The valuable mineral is mixed with lots of useless dirt and rock, which have to be separated off as much as possible. This is called concentrating the ore.

Some ores are already fairly concentrated when they are dug up – iron ore is often over 85% pure Fe_2O_3. But other ores are much less concentrated – copper ore usually contains less than 1% of the pure copper mineral.

Extracting metals: some of the issues

There is a range of factors to weigh up when thinking about the method for **extracting** a metal.

1 How can the ore be reduced?

The more reactive the metal, the harder it is to reduce its ore. The table on the right compares the methods used to reduce different ores.

2 Is there a good supply of ore?

Metals ores are mined in different parts of the world. If ore is not very pure, it may not be worth using – the cost of concentrating the ore may be too great. The more valuable the metal, the lower the quality of ore which can be used.

3 What are the energy costs?

It takes energy to extract metals, as well as a good supply of ore. This is especially true if the metal is extracted by electrolysis. For example, a quarter of the cost of making aluminium is the cost of electricity.

4 What is the impact on the environment?

Metals like iron and aluminium are produced on a huge scale. Millions of tonnes of ore are needed. Mining this ore can have a big environmental impact. This is why it is important to recycle metals. It takes about 250 kg of copper ore to make 1 kg of copper. So recycling 1 kg of copper means that 250 kg of ore need not be dug up.

MORE REACTIVE ↑ LESS REACTIVE

Metal	Method
potassium sodium calcium magnesium aluminium	electrolysis of molten ores
zinc iron tin lead	reduction of ores using carbon
silver gold	metals occur uncombined

Hot metal can be poured.

Questions

1 Suggest explanations for these facts:
   a The Romans used copper, iron, and gold, but not aluminium.
   b Iron is cheap compared with many other metals.
   c Gold is expensive, even though it is found uncombined in nature.
   d About half the iron we use is recycled, but nearly all the gold is recycled.
   e The tin mines in Cornwall have closed, even though there is still some tin ore left in the ground.

Key words
ores
oxides
reactive metals
extracting (a metal)

Extracting metals from ores

Zinc is a metal that can be extracted from its oxide. Zinc is found in the lithosphere as ZnS, called zinc blende. This can be easily turned to ZnO by heating it in air.

The task is to remove the oxygen from the zinc, to convert ZnO to Zn. Removing oxygen in this way is called **reduction**. The process needs a **reducing agent** which will remove oxygen. In removing oxygen, the reducing agent is **oxidized**.

Reducing zinc oxide to zinc using carbon

Carbon is often used as a reducing agent to extract metals. Carbon, in the form of coke, can be made cheaply from coal. At high temperatures, carbon has a strong tendency to react with oxygen, so it is a good reducing agent. What is more, the carbon monoxide formed is a gas, so it is not left behind to make the zinc impure.

How much metal?

Chemists often ask the 'How much?' question. It is useful to know, say, how much iron it is possible to get from 100 kg of pure iron ore, Fe_2O_3.

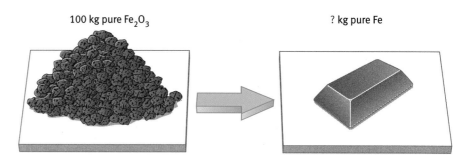

Relative atomic masses

Chemists need to know the relative masses of the atoms involved to answer questions such as 'How much Fe could you get from 100 kg of Fe_2O_3?'

Atoms are far too small to weigh directly. For example, it would take getting on for a million million million million hydrogen atoms to make one gram.

Key words

reduction
reducing agent
oxidized

Questions

2 a Write an equation for the reaction of zinc sulfide with oxygen to make zinc oxide and sulfur dioxide.

b What problems might arise from the formation of the sulfur dioxide on a large scale.

c What might be done to deal with this problem?

3 Why do oxidation and reduction always go together when carbon extracts a metal from a metal oxide?

Instead of working in grams, chemists find the mass of atoms relative to one another. Chemists can do this using an instrument called a mass spectrometer.

Values for the **relative atomic masses** of elements are shown in the periodic table on page 39. The relative mass of the lightest atom, hydrogen, is 1.

Formula masses

If you know the formula of a compound, you can work out its **relative formula mass** by adding up the relative atomic masses of all the atoms in the formula:

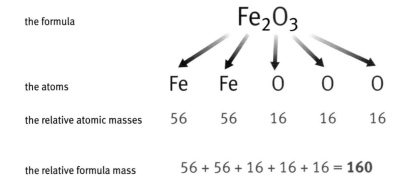

the formula Fe_2O_3

| the atoms | Fe | Fe | O | O | O |

the relative atomic masses   56   56   16   16   16

the relative formula mass $56 + 56 + 16 + 16 + 16 = \mathbf{160}$

Finding the formula mass of Fe_2O_3

Mg atoms weigh twice as much as C atoms.

Key words

relative atomic mass
relative formula mass

Example

How much Fe could you get from 100 kg of Fe_2O_3?

Fe_2O_3 has a relative formula mass of **160**

In this formula, there are 2 atoms of Fe
2Fe has relative mass $2 \times 56 = \mathbf{122}$

This means that, in 160 kg of Fe_2O_3, there must be 112 kg of Fe.

So 1 kg of Fe_2O_3 would contain 112/160 kg of Fe.

So 100 kg of Fe_2O_3 would contain 100 kg $\times$ 112/160 of Fe = 70 kg.

Another way of saying this is that the percentage of Fe in Fe_2O_3 is 70%.

Questions

Look up relative atomic masses in the periodic table on page 39.

4 What is the relative formula mass of carbon dioxide?

5 What mass of Al could be made from 1 tonne of Al_2O_3?

6 What mass of Na could be made from 2 tonnes of NaCl?

7 The main ore of chromium is $FeCr_2O_4$. What is the mass percentage of Cr in $FeCr_2O_4$?

8 1000 tonnes of copper ore are dug out of the ground. Only 1% of this is the pure mineral, $CuFeS_2$. What mass of the Cu could be made from 1000 tonnes of the ore?

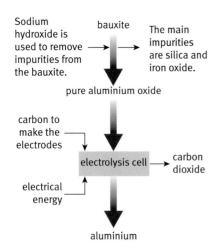

Sodium hydroxide is used to remove impurities from the bauxite.

bauxite

The main impurities are silica and iron oxide.

pure aluminium oxide

carbon to make the electrodes

electrolysis cell → carbon dioxide

electrical energy

aluminium

The processing of bauxite to aluminium

Extracting aluminium

Some reactive metals, such as aluminium, hold on to oxygen so strongly that they cannot be extracted using carbon as a reducing agent. To extract these metals, the industry has to use **electrolysis**.

Aluminium is the most abundant metal in the lithosphere. Much of the metal is in aluminosilicates. It is very hard to separate the aluminium from these minerals.

The main ore of aluminium is bauxite. This consists mainly of aluminium oxide, Al_2O_3. There is some iron in bauxite which has to be removed before extraction of aluminium.

The diagram below shows the equipment used to extract aluminium by **electrolysis**. The process takes place in steel tanks lined with carbon. The carbon lining is the negative **electrode**.

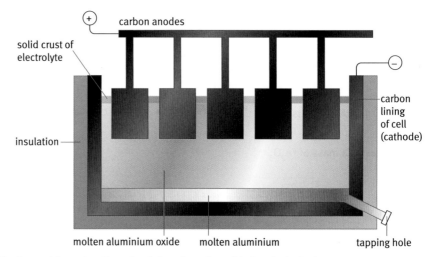

carbon anodes

solid crust of electrolyte

carbon lining of cell (cathode)

insulation

molten aluminium oxide   molten aluminium   tapping hole

Equipment for extracting aluminium from its oxide by electrolysis

The **electrolyte** is hot, molten Al_2O_3, which contains Al^{3+} and O^{2-} ions. Aluminium forms at the negative electrode. Aluminium is a liquid in the hot furnace and forms a pool of molten metal at the bottom of the tank.

The positive electrodes are blocks of carbon dipping into the molten oxide. Oxygen forms at the positive electrodes.

Ions into atoms and molecules

Electrolysis turns ions back into atoms. Metal ions are positively charged, so they are attracted to the negative electrode. It is a flow of electrons from the power supply into this electrode that makes it negative.

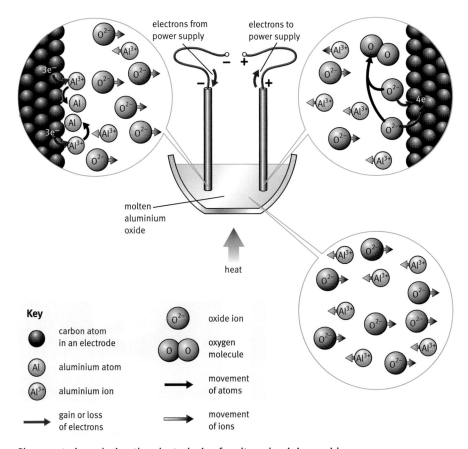

Changes to ions during the electrolysis of molten aluminium oxide

Positive metal ions gain electrons from the negative electrode and turn back into atoms. During the electrolysis of molten aluminium oxide, the aluminium ions turn into aluminium atoms:

$$Al^{3+} \quad + \quad 3e^- \quad \longrightarrow \quad Al$$

ion   electrons supplied by   atom
the negative electrode

Non-metal ions are negatively charged, so they are attracted to the positive electrode. This electrode is positive because electrons flow out of it to the power supply.

Negative ions give up electrons to the positive electrode and turn back into atoms. During the electrolysis of molten aluminium oxide, the oxide ions turn into oxygen atoms, which then pair up to make oxygen molecules:

$$O_2^- \quad \longrightarrow \quad O \quad + \quad 2e^-$$

ion   atom   electrons removed by
the positive electrode

$$O \quad + \quad O \quad \longrightarrow \quad O_2$$

atom   atom   molecule

Key words

electrolysis
electrode
electrolyte

Questions

9 Draw a diagram to show the electron arrangements in:
   a an aluminium atom
   b an aluminium ion

10 Sodium is extracted by the electrolysis of molten sodium chloride.
   a What are the two products of the process?
   b Use words and symbols to describe the changes at the electrodes during this process.

Find out about:

▶ the properties of metals

▶ the structure of metals

▶ bonding in metals

H Structure and bonding in metals

Metal properties

Metals have been part of human history for thousands of years. Our lives still depend on metals, despite the development of new materials, including all the different plastics. The varied uses of metals reflect their properties.

Technologists have learnt to use new metals and alloys so that, as well as steel and aluminium, other metals such as titanium and magnesium can now be used in engineering.

Many metals are strong. The titanium hull of this research submarine is strong enough to withstand the pressure at a depth of 6 km. Titanium is also used to make hip joints and racing cars.

Metals can be bent or pressed into shape. They bend without breaking. They are malleable. Aluminium sheet can be moulded under pressure to make cans.

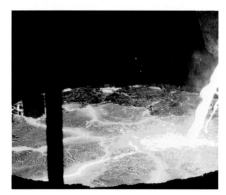

Most metals have high melting points.

Metals conduct electricity. Copper and aluminium are commonly used as conductors.

Metallic structures

Materials scientists use a model for the structure of metals which imagines that metal atoms are:

▶ tiny spheres

▶ arranged in a regular pattern

▶ packed close together in a crystal as a giant structure

Key words

metallic bonding

The diagram on the right shows the arrangement of atoms in copper, a typical metal. You can see how closely together the atoms of copper are packed. In fact, they are packed as close together as it is possible to be. Every atom has 12 other atoms touching it – the maximum number possible. The atoms are held together by strong metallic bonds. Because the bonds are strong, copper is strong and difficult to melt.

Metallic bonding

Metals have a special kind of bonding – not ionic, nor covalent, but metallic. **Metallic bonds** are strong, but flexible, so they allow the atoms to move to a new position.

Metal atoms tend to lose the electrons in their outer shell easily. In the solid metal, the atoms lose these electrons and become positive ions. The electrons, no longer held by the atoms, drift freely between the metal atoms, which are now positively charged. The attraction between the 'sea' of negative electrons and the positively charged metal atoms holds the structure together.

Overall, a metal crystal is not charged. This is because the total negative charge on the electrons balances the total positive charge on the metal ions.

The electrons can move freely between the ions, which explains why metals conduct electricity well. When an electric current flows through a metal wire, the free electrons drift from one end of the wire towards the other. Although the electrons are free, the metal ions themselves are packed closely together in a regular lattice.

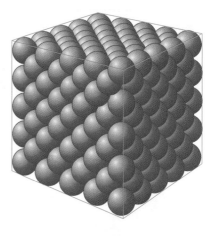

The arrangement of atoms in copper. Because the metallic bonds are strong, copper is strong and difficult to melt. Because the bonds are flexible, copper is malleable – the atoms can be moved around without shattering the structure.

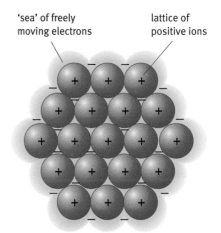

'sea' of freely moving electrons   lattice of positive ions

A model of metallic bonding

Questions

1. Give five examples of metals used for their strength. For each metal, give an example of a use that depends on the strength of the metal.

2. Someone looking at a model showing the arrangement of atoms in a copper crystal might think that the following statements are true. Which of these ideas are correct? Which ideas are false? What would you say to put someone right who believed the false ideas?
   a Copper crystals are shaped like cubes because the atoms are packed in a cubic pattern.
   b There is air between the atoms in a crystal of copper.
   c Copper is dense because the atoms are closely packed.
   d The atoms in a copper crystal are not moving at room temperature.
   e Copper has a high melting point because the atoms are strongly bonded in a giant structure.
   f Copper melts when strongly heated because the atoms melt

3. There are positive metal ions in a metal crystal, but a metal is not an ionic compound. Explain.

Find out about:

‣ impacts of extracting, using, and
disposing of metals

The life cycle of metals

Mining, mineral processing, and metal extraction produce many valuable metal products, but these activities can also have a serious impact on the environment. There can be a conflict between those that want to build up profitable industries and those whose aim is to protect the natural world.

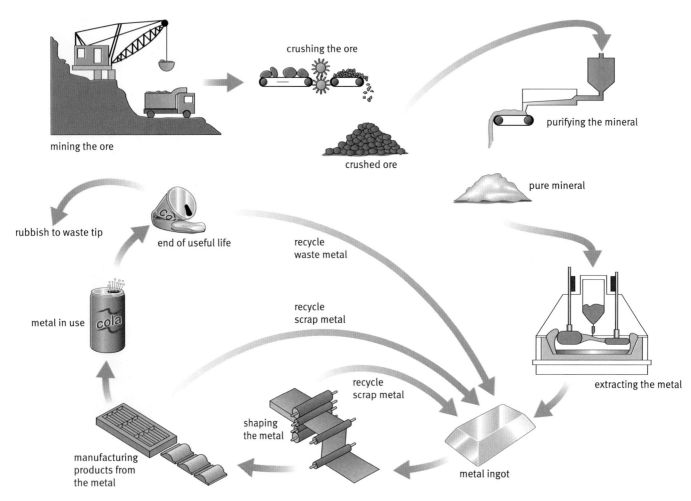

The life cycle of a metal article

Mining

Mining produces large volumes of waste rock and can leave very large holes in the ground. Miners use explosives to blast the rock. This is noisy and produces dust.

Open-cast mining for bauxite takes place in tropical countries such as Jamaica, Brazil, and Surinam. Separating the pure aluminium oxide leaves behind large volumes of red mud made up mostly of iron oxide. This has to be stored in large ponds where the very fine material can settle out.

Processing ores

Many metal ores are high value but low grade. The ore in an open-pit copper mine may contain as little as 0.4% of the metal and still be profitable. This means that 99.6% of the rock dug from the ground becomes waste. Near any mine there are waste tips. These can be a hazard if they contain traces of toxic metals such as lead or mercury.

Metal extraction

All the stages of metal extraction and metal fabrication need energy, use large volumes of water, and give off air pollutants. Higher expectations from society and tighter regulation mean that industries have to do more to prevent harmful chemicals escaping into the environment. Economic pressures favour the development of equipment and procedures that minimize the use of energy, water, and other resources.

Metals in use

Careful choice of metals can reduce the environmental impact of our life style. In transport, for example, lighter cars, trucks, and trains mean less fuel consumption and emissions, as well as less wear and tear on roads and tracks. Vehicles can be designed to be lighter by replacing steel with lighter metals such as aluminium and with plastics.

Recycling

Recycling is well established in the metal industries. Scrap metal from all stages of production is routinely recycled. Much metal is also recycled at the end of the useful life of metal products.

Recycled steel can be as good as new after reprocessing. The scrap is fed to a furnace and melted with fresh metal to make new steel. For every tonne of steel recycled, there is a saving of 1.5 tonnes of iron ore and half a tonne of coal. There is also a big reduction in the total volume of water needed, since large quantities of water are used in mineral processing.

Recycling aluminium is particularly cost-effective because so much energy is needed to extract the metal from its oxide by electrolysis. Recycling also reduces the impact on the environment by cutting the use of raw materials – and the associated mining and processing.

A large pond in Jamaica used to contain the waste from a mine to extract bauxite. Bauxite is impure aluminium oxide. The main impurity is iron oxide, which ends up in the rusty-looking waste.

Questions

1 Why are recycling rates for metal waste from manufacturing higher than recycling of metals after use?

2 According to an international company: 'Steel and aluminium products play a role in everything we do in modern-day life. So, although the production of steel and aluminium consumes resources and energy, both materials make a major contribution to our quality of life.' Draw up a table to list the benefits and costs of our use of these two metals. Do you agree with the claim made by this company?

3 The aluminium industry argues that aluminium is a sustainable material, because known bauxite reserves will last for hundreds of years at the current rates of production. Do you agree?

Summary

You have learnt how the theories of structure and bonding can help to explain the properties of the chemicals we find in the atmosphere, hydrosphere, biosphere and lithosphere. You have also learnt about some of the methods used to extract metals from their ores.

Molecules

▶ Most non-metal elements and most compounds between non-metal elements are molecular.

▶ The atoms in molecules are held together by strong covalent bonds.

▶ The attractive forces between molecules are weak, so that small molecules are often gases, such as the gases in the atmosphere.

▶ Water is a compound with small molecules that is a liquid which makes up most of the hydrosphere.

▶ Molecular compounds do not conduct electricity because their molecules are not charged.

▶ Living things are mainly made up from molecular compounds containing the elements carbon, hydrogen, oxygen, and nitrogen with small amounts of other elements.

▶ Carbohydrates, proteins, and DNA consist of long-chain molecules.

Giant ionic structures

▶ Compounds made of metals and non-metals have giant ionic structures.

▶ Ionic compounds have high melting points because of the strong attraction between the ions.

▶ Ionic compounds conduct electricity when molten or dissolved in water, when the ions are free to move.

▶ Salts are ionic compounds: some occur as minerals in the lithosphere; some dissolve in water and make the sea salty.

Giant covalent structures

▶ Silicon dioxide and diamond have giant covalent structures with atoms held together in a regular network with strong bonds.

▶ Chemicals with giant covalent structures have high melting points and do not dissolve in water.

▶ Giant structures do not conduct electricity because there are no free electrons or ions.

▶ Much of the lithosphere is made of giant covalent structures based on silicon, oxygen, and other elements, including aluminium.

Metallic structures

▶ All metal structures have a giant structure of metal atoms.

▶ The metallic bonding between the atoms is strong.

▶ Metals conduct electricity when solid and when molten because the bonding electrons are free to move.

▶ In a metal crystal there are positively charged ions held closely together by a sea of electrons that are free to move.

Ions into atoms

▶ Electrolysis turns ions back into atoms.

▶ Electrolysis splits an ionic compound and turns it back into its elements so it can be used to extract metals from ores.

▶ At a negative electrode, positive metal ions gain electrons and become metal atoms. H

▶ At a negative electrode, negative ions lose electrons and turn back into non-metal atoms (which may then join up to make molecules). H

Questions

1 A concept map is a web-like diagram for summarizing ideas and the links between them. Every concept map has 'nodes' which are boxes, with 'links' that are the lines between the boxes. Each node contains a concept. Each link includes a few words to show the relationships between the concepts. Arrow heads on the lines show the direction of relationship.

Create your own concept map for the topic of structure and bonding in chemicals of the environment. Include as many of the key words in this chapter as you can.

2 a Give two properties of a mineral that make it valued as a gemstone.

 b Diamond is a popular and valued gemstone.

 i Name the element in diamond.

 ii Why is diamond such a hard material?

 c An amethyst is a purple variety of quartz. It is a crystal of silicon dioxide with small amounts of iron impurity.

 i Why is amethyst cheaper than diamond?

 ii Why does amethyst have a very high melting point?

3 Passing an electric current through molten calcium bromide produces a metallic bead at one electrode and a red-brown gas at the other. Calcium bromide consists of calcium ions, Ca^{2+}, and bromide ions, Br^- The metallic bead reacts with water to produce a gas that burns with a squeaky pop.

 a What is the name for the process of splitting a compound with an electric current?

 b i Identify the metal formed and the red-brown gas.

 ii Which of the products forms at the negative electrode?

 c Draw and label a diagram of an apparatus that could be used to pass an electric current through molten calcium bromide.

 d i Why does solid calcium bromide not conduct electricity?

 ii Write the chemical formula of calcium bromide.

 e Explain the changes when the current flows at the electrode that produces the metallic bead.

4 This is one of the reactions for extraction iron from its ore in a furnace:

$$Fe_2O_3(s) + 3CO(g) \rightarrow 2Fe(s) + 3CO_2(g)$$

 a In this reaction which chemical is:

 i oxidized?

 ii reduced?

 iii the reducing agent?

 b What mass of iron can be obtained from 16 tonnes of iron oxide? (See page 39 for the relative atomic masses.)

 c The carbon monoxide comes from coke (C) in the furnace. What mass of coke is needed to form the carbon monoxide required to extract the metal from 16 tonnes of iron oxide?

4 Aluminium is a more reactive metal than iron. It can be extracted from its oxide, Al_2O_3, only by electrolysis.

 a Name one other metal that can only be extracted by electrolysis.

 b Write an equation to show what happens at the negative electrode during the electrolysis of molten aluminium oxide.

 c Calculate the mass of aluminium that can be extracted from 1000 kg of aluminium oxide.

5 Make a list of five metals in common use.

 a For each metal:

 i give an example of how it is used

 ii state a property of the metal that it makes it more suitable for that use than the other metals in your list

 b Explain why all metals conduct electricity.

Why study electric circuits?

Imagine life without electricity – rooms lit by candles or oil lamps, no electric cookers or kettles, no radio, television, computers, or mobile phones, no cars or aeroplanes. Electricity has transformed our lives, but you need to know enough to use it safely. More fundamentally, electric charge is one of the basic properties of matter – so anyone who wants to understand the natural world around them needs to have some understanding of electricity.

The science

The particles of which atoms are made carry an electric charge. An electric current is a flow of charges. A useful model of an electric circuit is to imagine the wires full of charges, being made to move around together by the battery. The size of current depends on the battery voltage and the resistance of the circuit. A voltage can also be produced by moving a magnet near a coil. This is used to generate electricity on a large scale.

Physics in action

The scientific understanding of electricity was developed over quite a short period, from about 1800 to 1840. Nowadays scientists use electricity, or instruments that depend on electricity, in almost every aspect of their work. One important focus of research in the 21st century is on the development of new ways of generating electricity, using renewable energy sources such as sunlight, wind, and waves.

Electric circuits

Find out about:

▶ the idea of electric charge, and how moving charges result in an electric current

▶ how models that help us 'picture' what is going on in an electric circuit can be used to explain and predict circuit behaviour

▶ electric current, voltage, and resistance

▶ energy transfers in electric circuits, and how mains electricity is generated and distributed

Small sparks caused by static electricity are harmless, but on a larger scale they can be much more dangerous. Lightning is an electrical spark between a thundercloud and the ground.

A charged comb attracts a stream of water.

A Static electricity

When you get out of a car, you sometimes get a small electric shock when you touch the metal door – and you might hear a little 'crack' as a spark jumps between your hand and the car door. You sometimes hear the same sort of little crackles when you take off a jumper. In a dark room, you may be able to see the sparks. They are caused by **static electricity**. Electricity is part of our everyday world.

Charging by rubbing

Electrical effects can be produced by rubbing two materials together. If you rub a balloon against your jumper, the balloon will stick to a wall. If you rub a plastic comb on your sleeve, the comb will pick up small pieces of tissue paper – they are attracted to the comb. In both cases, the effect wears off after a short time.

When you rub a piece of plastic, it is somehow changed: it can then affect objects nearby. The more it is rubbed, the stronger the effect. It seems that something is being stored on the plastic. If a lot is stored, it may escape by jumping to a nearby object, in the form of a spark. We say that the plastic has been charged.

Charging by rubbing also explains the examples discussed at the top of the page. When you get out of a car, you slide across the seat, rubbing your clothes against it. As you pull off a jumper, it rubs against the shirt or blouse you are wearing beneath it.

Two types of charge

If you rub two identical plastic rods and then hold them close together, the rods push each other apart – they **repel**. The forces they exert on each other are very small, so you can only see the effect if one of the rods can move easily.

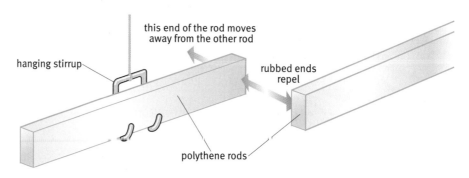

Two rubbed polythene rods repel each other. The hanging rod moves away from the other one.

If you try this with two rods of different plastics, however, you can find some pairs that **attract** each other. Scientists' explanation of this is that there are two types of **electric charge**. If two rods have the same type of charge, they repel each other. But if they have charges of different types, they attract. The early electrical experimenters called the two types of charge **positive** and **negative**. These names are just labels. They could have called them red and blue, or A and B.

Where does charge come from?

Scientists believe that charge is not *made* but is *moved around* when two things are rubbed together. If you rub a plastic rod with a cloth, both the rod and the cloth become charged. (To see this, you need to wear a polythene glove on the hand holding the cloth, otherwise the charge will escape through your body.) Each object gets a different charge: if the rod has a positive charge, the cloth has a negative one. Rubbing does not make charge. It separates charges that were there all along.

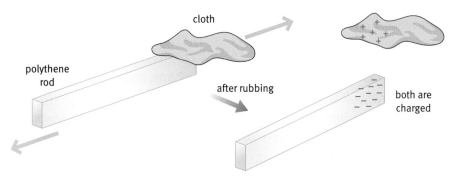

After it has been rubbed, the rod has a negative charge and the cloth has a positive charge. A possible explanation is that some electrons have been transferred from the cloth to the rod.

Questions

1 Imagine that you have two plastic rods which you know get a positive and a negative charge when you rub them with a cloth. Now you are given a third plastic rod. Explain how you could test whether it gets a positive or a negative charge when rubbed.

2 Some picture frames are made with plastic rather than glass. If you clean the plastic with a duster, it may get dusty again very quickly. Use the ideas on these pages to explain why this happens.

Attracting light objects

An object with a positive charge attracts another object with a negative charge. But why does a charged rod also attract light objects, such as little pieces of paper? The reason is that there are charges in the paper itself, all the time. Normally these are mixed up together, with equal amounts of each. So a piece of paper is uncharged. If a negatively charged rod comes near, it repels negative charges in the paper to the end farthest away. This leaves a surplus of positive charges at the near end. The attraction between these positive charges and the rod is stronger than the repulsion between the negative charges (at the far end) and the rod. So the little piece of paper is attracted to the rod.

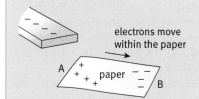

A charged rod separates the charges in a piece of paper nearby – and the paper is attracted to the rod.

electric field lines

The lines around this positively charged ball are one way of showing the electric field. The field is strongest where the lines are close. Another positive charge entering the field will feel a force in the direction of the arrows.

What is electric charge?

Charge is a basic property of matter, which cannot be explained in terms of anything simpler. All matter is made of atoms, which in turn are made out of protons (positive charge), neutrons (no charge), and **electrons** (negative charge). In most materials there are equal numbers of positive and negative charges, so the whole thing is neutral. When you charge something, you move some electrons to it or from it.

Chemists think of the atom as a tiny nucleus, with a positive charge, surrounded by a cloud of electrons, which have negative charge. As the electrons are on the outside, they can be 'rubbed off', on to another object.

Although they cannot explain charge, scientists have developed useful ideas for predicting its effects. An example is the idea of an **electric field**. Around every charge there is an electric field. In this region of space, the effects of the charge can be felt. Another charge entering the field will experience a force.

Moving charge = current

When a van de Graaff generator is running, charge collects on its dome. As the charge builds up, the electric field around the dome gets stronger. The charge may 'jump' to another nearby object, in the form of a spark. If you hold a mains-testing screwdriver close to the dome and touch its metal end-cap, the indicator lamp inside the screwdriver lights up. So there is an **electric current** through the lamp, making it light. Charge on the dome is escaping across the air gap, through the indicator lamp, and through you to the Earth. This (and other similar observations) suggests that an electric current is a flow of charge.

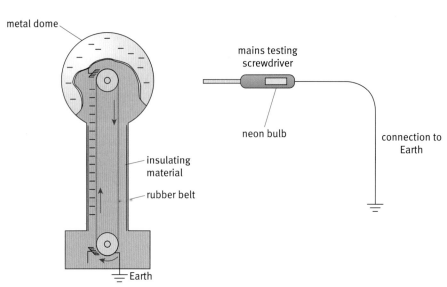

As charge flows from the dome, through the indicator lamp, it makes it light up. This suggests that an electric current is a flow of charge.

Key words

static electricity
electric charge
repel, attract
positive, negative
electron
electric field
electric current

B Simple circuits

A closed loop

The diagram below shows a simple **electric circuit**. If you make a circuit like this, you can quickly show that:

▶ If you make a break *anywhere* in the circuit, *everything* stops.

When you open the switch in this electric circuit, both bulbs go off.

This suggests that something has to go all the way round an electric circuit to make it work. This 'something' is electric charges. If it was enough for the charges simply to go from the battery to the lamp, then one of the lamps would be lit, even with the switch open. But this does not happen. There has to be a complete loop – from one terminal of the battery, through the lamps and switch, and back to the other battery terminal.

You will also notice that:

▶ Both lamps come on *immediately* when the circuit is completed. And they go off *immediately* if you make a break in the circuit.

Perhaps this is because the circuit is small. So imagine making a 'big circuit' with much longer wires than usual. When you turn on the switch, it is still impossible to see any delay before the lamp lights. So the size of the circuit makes no difference. Remember that an electric current is moving charges. So perhaps charges move very quickly through the wires, from the battery to the lamp, as soon as the circuit is switched on. However, there is a better explanation. Imagine that there are charges in all the components of the circuit (wires, lamp filaments, batteries) *all the time*. Closing the switch just allows these charges to move. They all move together, so the effect is immediate, even though the charges themselves do not move very fast.

Find out about:

▶ how simple electric circuits work
▶ models that help explain and predict the behaviour of electric circuits

When you unscrew one bulb, you make a break in the circuit and all the lights go out.

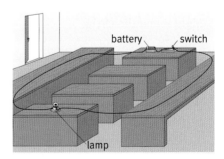

A 'big circuit' with longer wires than usual. When we switch on, the bulb lights immediately. There is no delay.

153

Some ways of thinking about charges in an electric circuit

Peas in a pipe

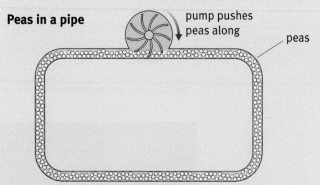

Imagine a plastic pipe full of dried peas, formed into a closed loop, with a pump that can push the peas along. Because the peas are close together, when one moves, they all move. The effect is immediate. If a barrier is inserted anywhere, it will stop them moving everywhere – instantly.

Moving rope role-play

Here is another way to think about an electric circuit. Imagine a group of about six to eight people standing in a circle, about 1 metre apart. Each person holds out both hands, palms upwards and fingers slightly bent. A continuous loop of rope passes from person to person, right round the circle.

Chain on a bike

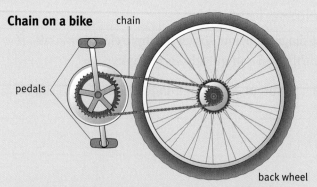

When you turn the pedals of a bike, this makes the chain move. When one part of the chain moves, it all moves. So it immediately makes the back wheel turn. There is no delay. The effect of your pedalling is transmitted immediately to the back wheel.

One person now acts as the 'battery'. She makes the rope slowly move round, passing through everyone's hands as it goes. Notice that the effect is instantaneous. As soon as the rope moves anywhere, it moves everywhere. If everyone keeps this going for a short time, they will notice two things happening. First, the 'battery' begins to tire – because she is transferring stored energy from her muscles to make the rope move. Second, the hands of the people around the circle begin to get slightly hot – because of the friction force between their hands and the rope. The battery is doing work against this friction force, and this is causing some heating.

This is a very similar to an electric circuit. The battery makes current move round the circuit – and this does work on all the other components round the circuit.

Questions

1 Imagine making the circuit at the top of page 153, but with the switch between the two batteries instead of between the two lamps. Explain what would now happen when the switch is opened and closed.

2 How do the three models above help to explain the following observations about electric circuits?
   a When you switch a circuit on, any bulbs and motors in it come on immediately, no matter where they are in the circuit:
   b in a simple circuit, a switch anywhere starts and stops everything.

3 In the electric circuit role-play, how could you illustrate the effect of putting a second battery into a circuit?
   a pointing in the same direction as the first;
   b pointing in the opposite direction to the first?
   Does the role-play situation correctly predict what happens in a real circuit?

An electric circuit model

A model is a way of thinking about how something works, under the surface. The diagram on the right summarizes the scientific model of an electric curcuit. The key ideas are:

▶ Charges are present throughout the circuit all the time.

▶ When the circuit is a closed loop, the battery makes the charges move.

▶ All of the charges move round together.

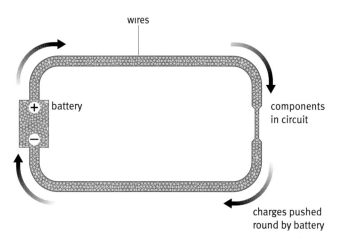

You can think of an electric circuit as a flow of **charges**, which are present in all materials (and free to move in conductors), moving round a closed conducting loop, pushed by the battery.

The battery makes the charges move in the following way. Chemical reactions inside the battery have the effect of separating electric charges, so that positive charge collects on one terminal of the battery and negative charge on the other. If the battery is connected into a circuit, the charges on the battery terminals set up an electric field in the wires of the circuit. This makes free charges in the wire drift slowly along. However, even though the charges move slowly, they all begin to move at once, as soon as the battery is connected. So the effect of their motion is immediate. Notice too that the flow of charge is continuous, all round the circuit. Charge also flows through the battery itself.

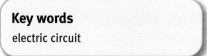

Key words
electric circuit

Conventional current, electron flow

In the model above, the charges in the circuit are shown moving away from the positive terminal of the battery, through the wires and other components, and back to the negative terminal of the battery. This assumes that the moving charges are positive. In fact there is no simple way of telling whether the moving charges are positive or negative, or which way they are moving. Several decades after this model was first proposed and had become generally accepted, scientists came to realize that the moving charges in metals were electrons, which have negative charge. In all metals, the atoms have some electrons that are only loosely attached to their 'parent' atom and are relatively free to wander through the metal. It is these that move, in the weak electric field that the battery sets up in the wire.

To explain and predict how electric circuits behave, it makes no difference whether you think in terms of a flow of electrons in one direction or positive charges in the other. Although scientists now believe it is electrons that flow in metals, in this course use the model of conventional current going the other way, as most physicists and engineers do.

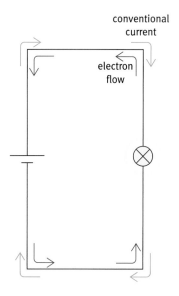

An electric current is a flow of electrons through the wires of the circuit. You can think of it equally well as a 'conventional current' of positive charges going the other way.

Find out about

▶ how to measure electric current
▶ the size of the electric current at different places in a circuit

Typical currents

Currents smaller than 1 amp are often measured in milliamps (mA) or microamps (μA).

$1 A = 1000 mA = 1\,000\,000\,μA$

Typical size of the current in some common applications:

in an electric kettle: 8 A

in a torch bulb: 0.3 A or 300 mA

in a radio: 0.1 A or 100 mA

in a calculator: 0.005 A or 5 mA

in a digital watch: 0.000 05 A or 50 μA

c Electric current

An electric current is a flow of charge. You cannot see a current, but you can observe its effects. The current through a torch bulb makes the fine wire of the filament heat up and glow. The bigger the current through a bulb, the brighter it glows (unless, the current gets too big and the bulb 'blows'). So the brightness of a bulb shows the size of the current through it.

A better way to measure the size of an electric current is to use an **ammeter**. The reading (in amperes, or amps (A) for short) indicates the amount of charge going through the ammeter every second.

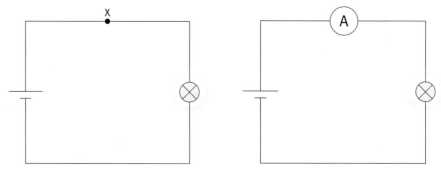

To measure the current at point X, you have to make a gap in the circuit at X and insert the ammeter in the gap, so that the current flows through it.

Current around a circuit

If you use an ammeter to measure the size of the electric current at different points around a circuit, you get a very important result.

▶ The current is the same everywhere in a simple (single-loop) electric circuit.

This may seem surprising. Surely the bulbs must use up current to light. But this is not the case. Current is the movement of charges in the wire, all moving round together like dried peas in a tube, or a moving belt or chain. So the current at every point round the circuit must be the same.

Of course, *something* is being used up. It is the energy stored in the battery. This is getting less all the time. The battery is doing work to push the current through the filaments of the light bulbs, and this heats them up. The light then carries energy away from the glowing filament. So the circuit is transferring energy from the battery, to the bulb filaments, and then on to the surroundings (as light). The current enables this energy transfer to happen. But the current itself is not used up.

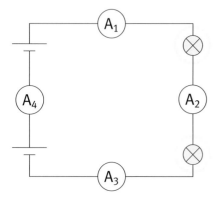

The current is the same size at all these points – even between the batteries. Current is not used up to make the bulbs light.

Branching circuits

Often, you want to run more than one thing from the same battery. One way to do this is to put them all in a single loop, one after the other. Components connected like this are said to be **in series**. The moving charges then have to pass through each of them in turn.

Another way is to connect components **in parallel**. In the circuit on the right, the two bulbs are connected in parallel. This has the advantage that each bulb now works independently of the other. If one burns out, the other will stay lit. This makes it easy to spot a broken one and replace it.

Currents in the branches

Look at the circuit below, which has a motor and a buzzer connected in parallel. A student, Nicola, was asked to measure the current at points a, b, c, and d. Her results are shown in the table below on the right.

In this portable MP3 player, the battery has to run the motor that turns the hard drive, the head that reads the disk, and the circuits that decode and amplify the signals.

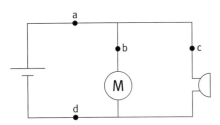

Point in circuit	Current (mA)
a	230
b	150
c	80
d	230

Measuring currents in a circuit with two parallel branches.

Nicola noticed that the current is the same size at points a and d: 230 mA. When she added the currents at b and c, the result was also 230 mA. This makes good sense if you think about the model of charges moving round. At the junctions, the current splits, with some charges flowing through one branch and the rest flowing through the other branch. Current is the amount of charge passing a point every second. So the amounts in the two branches must add up to equal the total amount in the single wire before or after the branching point.

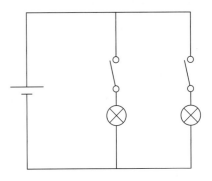

An advantage of connecting components in parallel is that each can be switched on and off independently.

> **Key words**
> ammeter
> in series
> in parallel

> **Questions**
>
> 1 Look at the circuit at the top of page 156. Draw a circuit diagram showing how you would connect an ammeter to measure the current in the wire between the bulb and the negative terminal of the battery.
>
> 2 When we want to run several things from the same battery, it is much more common to connect them in parallel than in series. Write down three advantages of parallel connections.
>
> 3 Look at the circuit above on the right. If you wanted to switch both bulbs on and off together, where would you put the switch? Draw two diagrams showing two possible positions of the switch that would do this.
>
> 4 In the circuit above with the motor and buzzer, what size is the electric current:
>   a in the wire just below the motor?
>   b in the wire just below the buzzer?
>   c through the battery itself?

Find out about:

▶ how the battery voltage and the circuit resistance together control the size of the current

▶ what causes resistance

▶ the links between battery voltage, resistance, and current

All the batteries on the front row are marked 1.5 V – but are very different sizes. The three at the back are marked 4.5 V, 6 V, and 9 V

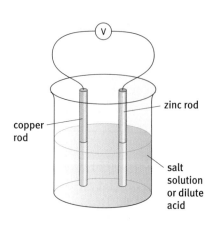

A simple battery. The voltage depends on the metals and the solution you choose.

D Controlling the current

The scientific model of an electric circuit imagines a flow of charge round a closed conducting loop, pushed by a battery. The next step is to ask how the size of the current can be controlled. As the current in a circuit is caused by the battery, this is a good place to begin.

Battery voltage

Batteries come in different shapes and sizes. They usually have a **voltage** measured in volts (V), marked clearly on them, for example, 1.5 V, 4.5 V, 9 V. To understand what voltage means and what this number tells you, look at the following diagrams, which show the same bulb connected first to a 4.5 V battery and then to a 1.5 V battery.

With a 4.5 V battery, this bulb is brightly lit.

With a 1.5 V battery, the same bulb is lit, but very dimly.

The bigger the current through a light bulb, the brighter it will be (up to the point where it 'blows'). So the current through the bulb above is bigger with the 4.5 V battery. You can think of the voltage of a battery as a measure of the 'push' it exerts on the charges in the circuit, or the amount of work it does pushing charges round the circuit. The battery sets up an electric field in the wires of the circuit, and this makes the free charges move round. The bigger the voltage, the bigger the 'push' – and the bigger the current as a result.

The battery voltage depends on the choice of chemicals inside it. To make a simple battery, all you need are two pieces of different metal and a beaker of salt solution or acid. The voltage quickly drops, however. The chemicals used in real batteries are chosen to provide a steady voltage for several hours of use.

Resistance

The size of the current in a circuit depends on the battery voltage, but this is not the only thing that matters. The components in the circuit provide a **resistance** to the flow of charge. The battery 'pushes' against this resistance. You can see the effect of this if you compare two circuits with different resistors. Resistors are components designed to control the flow of charge.

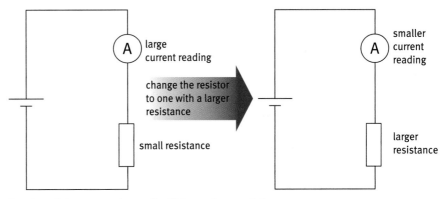

The size of the current is smaller if the resistance is larger

Changing resistance changes the size of the current. The bigger the resistance, the smaller the current.

What causes resistance?

Everything has resistance, not just special components called resistors. The resistance of connecting wires is very small, but not zero. Other kinds of metal wire have larger resistance. The filament of a light bulb has a lot of resistance. This is why it gets so hot when there is a current through it. A heating element, like that in an electric kettle, is just a resistor.

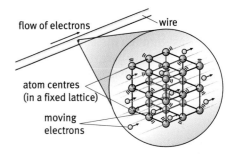

Why the temperature of a wire rises when a current flows through it

All metals get hot when charge flows through them. In metals, the moving charges are free electrons. As they move round, they collide with the fixed array, or lattice, of atoms in the wire. These collisions make the atoms vibrate a little more, so the temperature of the wire rises. In some metals, the fixed atoms provide only small targets for the electrons, which can get past them relatively easily. In other metals, the fixed atoms present a much bigger obstacle – and so the resistance is bigger.

Key relationships in an electric circuit: a summary

The size of the electric current (I) in a circuit depends on the battery voltage (V) and the resistance (R) of the circuit.

▶ If you make V bigger, the current (I) increases.

▶ If you make R bigger, the current (I) decreases.

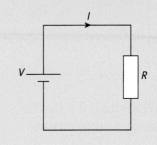

Questions

1 Look back at the moving rope role-play described on page 154. How would you role-play an increase in resistance? What effect would this have? (You should be able to think of at least two.) Explain how the role-play helps to predict the behaviour of a real circuit.

2 Suggest two different ways in which you could change a simple electric circuit to make the electric current bigger.

Ohm's law

Ohm's law says that the current through a conductor is proportional to the voltage across it – provided its temperature is constant.

It applies only to some types of conductor (such as metals).

An electric current itself causes heating, which complicates matters. For example, the current through a light bulb is not proportional to the battery voltage. The *I–V* graph is curved. The reason is that the current through the bulb filament heats it up – and its resistance increases with temperature.

Questions

3 In Keiko's investigation:

a How many 1.5 V batteries would she need to use to make a current of 600 mA flow through her coil?

b What is the resistance of her coil, in ohms?

4 In the circuit below, a 9 V battery is connected to a 45 Ω resistor.

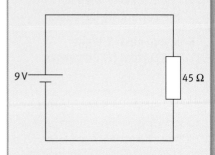

What size is the electric current in the circuit?

Measuring resistance

You can explore the relationship between battery voltage and current in more detail. Keiko did this by measuring the current through a coil of wire with different batteries.

1 Keiko connected a coil of resistance wire to a 1.5 V battery and an ammeter. She noted the current.

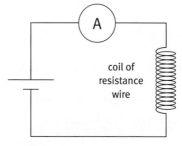

coil of resistance wire

2 She then added a second battery in series. She noted the current again.

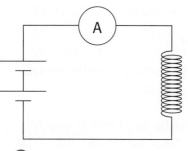

3 Keiko repeated this with 3, 4, 5, and 6 batteries, to get a set of results:

4 Finally, she drew a graph of current against battery voltage:

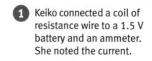

Number of 1.5 V batteries	Battery voltage (V)	Current (mA)
1	1.5	75
2	3.0	150
3	4.5	225
4	6.0	300
5	7.5	375
6	9.0	450

The straight-line graph means that the current in the circuit is proportional to the battery voltage. This result is known as **Ohm's law**. The number you get if you divide voltage by current is the same every time. The bigger the number, the larger the resistance. This is how to measure resistance:

$$\text{resistance of a conductor} = \frac{\text{voltage across the conductor}}{\text{current through the conductor}}$$

$$R = \frac{V}{I}$$

The units of resistance are called ohms (Ω).

Rearranging this equation gives $I = V/R$. You can use this to calculate the current in a circuit, if you know the battery voltage and the resistance of the circuit.

Variable resistors

Resistors are used in electric circuits is to control the size of the current. Sometimes we want to be able to vary the current easily, for example to change the volume on a radio or CD player. A variable resistor is used. This is a resistance whose size can be steadily changed by turning a dial or moving a slider.

The circuit diagram on the right shows the symbol for a variable resistor. As you alter its resistance, the brightness of the light bulb changes, and the readings on *both* ammeters increase and decrease together. The variable resistor controls the size of the current everywhere round the circuit loop.

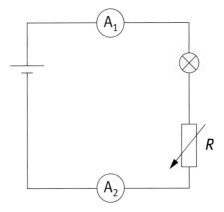

Using a variable resistor to control the current in a series circuit

Each of these sliders adjusts the value of a variable resistor.

Some useful sensing devices are really variable resistors. For example, a light-dependent resistor (LDR) is a semiconductor device whose resistance is larger in the dark but gets smaller as the light falling on it gets brighter. An LDR can be used to measure the brightness of light or to switch another device on and off when the brightness of the light changes. For example, it could be used to switch an outdoor light on in the evening and off again in the morning.

A thermistor is another device made from semiconductor material. Its resistance changes rapidly with temperature. The commonest type has a lower resistance when it is hotter. Thermistors can be used to make thermometers (to measure temperature) or to switch another device on or off as temperature changes. For example, a thermistor could be used to switch an immersion heater on when the temperature of water in a tank falls below a certain value and off again when the water is back at the required temperature.

> **Key words**
> voltage
> resistance
> Ohm's law

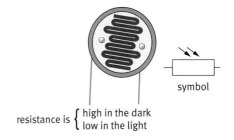

resistance is { high in the dark / low in the light

A light-dependent resistor (LDR)

One type of thermistor

161

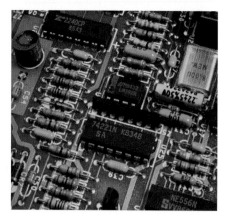

This circuit board from a computer contains a complex circuit, with many components. The small cylindrical ones are resistors.

Combinations of resistors

Most electric circuits are more complicated than the ones discussed so far in this chapter. Circuits, usually contain many components, connected in different ways. There are just two basic ways of connecting circuit components: in series or in parallel.

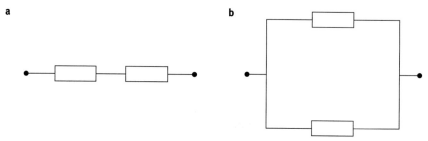

Two resistors connected **a** in series, **b** in parallel

Two resistors in series have a larger resistance than one on its own. The battery has to push the current through both of them. But connecting two resistors in parallel makes a smaller total resistance. There are now two paths that the moving charges can follow. Adding a second resistor in parallel does not affect the original path but adds a second equivalent one. It is now easier for the battery to push charges round, so the resistance is less.

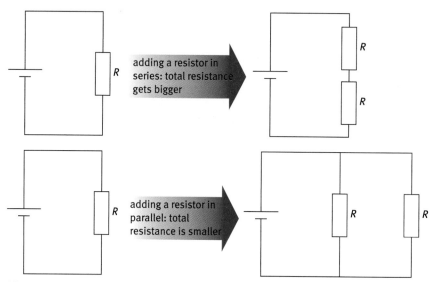

Different ways of adding a second resistor: how do you do it makes a difference

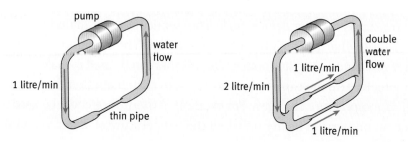

A water flow model shows how the total resistance gets less when a second parallel path is added.

Questions

5 All the resistors in the three diagrams below are identical. Put the groups of resistors in order, from the one with the largest total resistance to the one with the smallest total resistance.

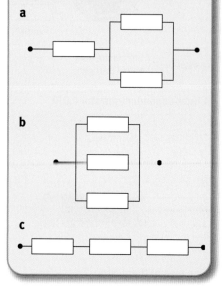

E Potential difference

You can think of the voltage of a battery as a measure of the 'push' it exerts on the charges in a circuit. But a voltmeter also shows a reading if you connect it across any resistor (or bulb) in a working circuit. Resistors and bulbs do not 'push'. So the **voltmeter** reading must be indicating something else.

A useful picture is to think of the battery as a pump, lifting water up to a higher level. The water then drops back to its original level as it flows back to the inlet of the pump. The diagram below shows how this would work for a series circuit with three resistors (or three lamps). The pump increases the potential energy of the water. The water then loses this energy in three steps. The total amount of energy lost has to be equal to the amount of energy gained.

In the electric circuit, the battery does work on the electric charges, to lift them up to a higher 'energy level'. They then transfer energy in three stages as they drop back to their starting level. A voltmeter measures the difference in 'level' between the two points it is connected to. This is called the potential difference between these points. **Potential difference (p.d.)** is measured in volts (V).

Find out about:
▶ how voltmeters measure the potential difference between two points in a circuit
▶ how height provides a useful model for thinking about electrical potential
▶ how current splits between parallel branches

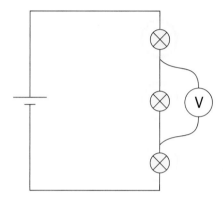

There is a reading on this voltmeter. This cannot be a measure of the strength of a 'push' – so what is it telling us?

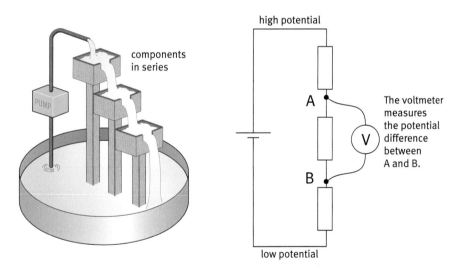

components in series

high potential

A

B

The voltmeter measures the potential difference between A and B.

low potential

The voltage of a battery is the potential difference between its terminals. If you put a battery with a larger voltage into the circuit above, this would mean a bigger potential difference across its terminals. The potential difference across each lamp (or resistor) would also now be bigger. Going back to the water pump model, this is like changing to a stronger pump that lifts the water up to a higher level. The three downhill steps then also have to be bigger, so that the water ends up back at its starting level.

There is a potential difference of 12 V across the terminals of a car battery.

The same idea works for a parallel circuit. In this case, the water divides into three streams. Each loses all its energy in a single step.

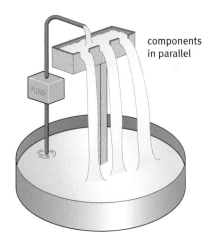

components in parallel

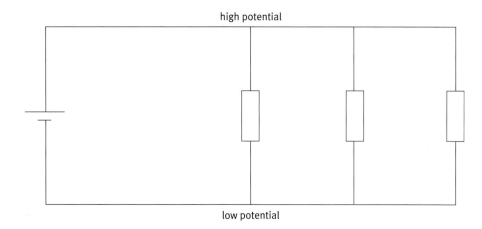

Voltmeter readings across circuit components

This water pump model helps to explain and to predict voltmeter readings across resistors in different circuits. If several resistors are connected in parallel to a battery, the potential difference across each is the same. It is equal to the battery voltage.

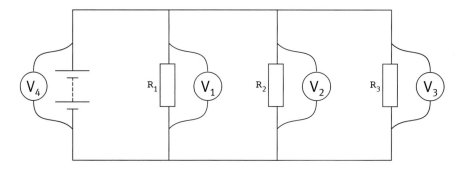

All of these voltmeters will have the same reading, even if R_1, R_2, and R_3 are different.

If the resistors are connected in series (as on the left), the sum of the p.d.s across them is equal to the battery voltage. This is exactly what you would expect from the 'waterfall' picture on page 163.

In the series circuit, the p.d. across each resistor depends on its resistance. The biggest voltmeter reading is across the resistor with biggest resistance. Again this makes sense. More work has to be done to push charge through a big resistance than a smaller one.

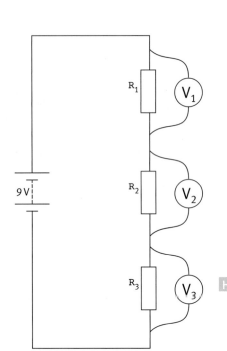

The voltages are in proportion to the resistances. And their sum is equal to the battery voltage.

Currents in parallel branches

The potential difference across resistors R_1, R_2, and R_3 in the parallel circuit the opposite page is exactly the same for each. It is equal to the p.d. across the battery itself. But the currents through the resistors are not necessarily the same. This will depend on their resistances. The current through the biggest resistor will be the smallest. There are two ways to think of this:

1 Imagine water flowing through a large pipe. The pipe then splits in two, before joining up again later. If the two parallel pipes have different diameters, more water will flow every second through the pipe with the larger diameter. The wider pipe has less resistance than the narrower pipe to the flow of water So the current through it is larger.

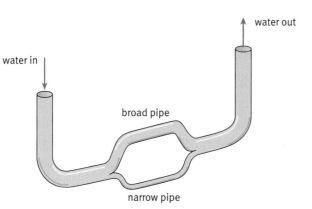

More water flows each second through the larger pipe. It has less resistance to the water flow.

2 Think of two resistors connected in parallel to a battery as making two separate simple loop circuits that share the same battery. The current in each loop is independent of the other. The smaller the resistance in a loop, the bigger the current. Some wires in the circuit are part of both loops, so here the current will be biggest. The current here will be the sum of the currents in the loops.

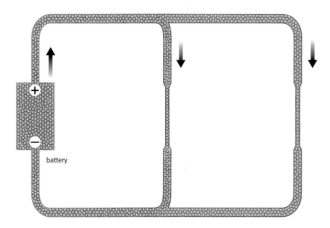

A parallel circuit like this behaves like two separate simple loop circuits.

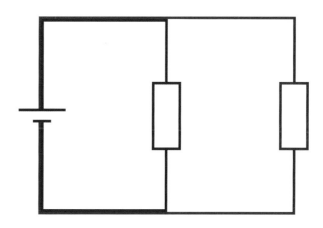

Key words

voltmeter
potential difference (p.d.)

Questions

1 Imagine removing the red resistor from the circuit below leaving a gap. What would happen to:
  a the current through the purple resistor?
  b the current from (and back to) the battery?
  Explain your reasoning each time.

Find out about:

▶ how the power produced in a circuit component depends on both current and voltage

F Electrical power

An electric circuit is primarily a device for doing work of some kind. It transfers energy initially stored in the battery to somewhere else. A key feature of any electric circuit is the rate at which work is done on the components in the circuit – that is, the rate at which energy is transferred from the battery to the other components. This is called the **power** of the circuit.

Measuring the power of an electric circuit

Imagine starting with a simple battery and bulb circuit and trying to double, and treble, the power. You could do this in two ways.

▶ One is to add a second bulb, and then a third, in parallel with the first. In the circuits down the left-hand side of the diagram below, the p.d. is the same, but the current supplied by the battery doubles and trebles. The power is proportional to the current.

▶ Another is to add a second bulb, and then a third, in series with the first. Now you need to add a second battery, and then a third, to keep the brightness of the bulbs the same each time. In these circuits (across the diagram below), the current is the same, but the p.d. of the battery doubles and trebles. The power is proportional to the current.

This is summarized in the box in the bottom right-hand corner.

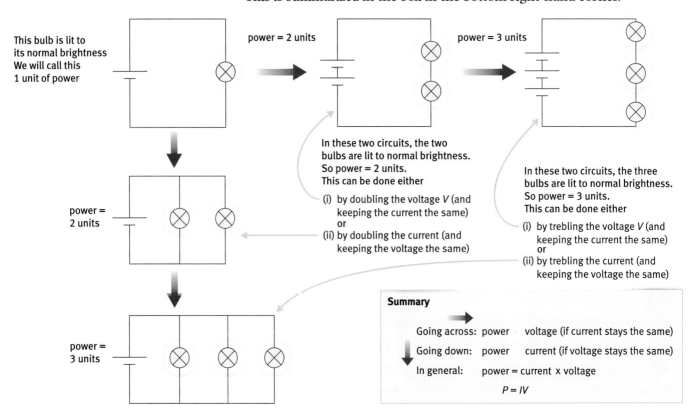

This bulb is lit to its normal brightness We will call this 1 unit of power

power = 2 units

power = 3 units

power = 2 units

power = 3 units

In these two circuits, the two bulbs are lit to normal brightness. So power = 2 units. This can be done either

(i) by doubling the voltage *V* (and keeping the current the same) or

(ii) by doubling the current (and keeping the voltage the same)

In these two circuits, the three bulbs are lit to normal brightness. So power = 3 units. This can be done either

(i) by trebling the voltage *V* (and keeping the current the same) or

(ii) by trebling the current (and keeping the voltage the same)

Summary

Going across: power   voltage (if current stays the same)

Going down: power   current (if voltage stays the same)

In general: power = current × voltage

$$P = IV$$

In general, the power dissipated in an electric circuit depends on both the current and the voltage:

power	=	current	×	voltage
P	=	I		V
(watt, W)		(ampere, A)		(volt, V)

The unit of power is the watt (W). One watt is equal to one joule per second. So if you know the power, it is easy to calculate how much work is done (or how much energy is transferred) in a given period of time:

work done (or energy transferred)	=	power	×	time
(joule, J)		(watt, W)		(second, s)

H To see how this equation for power makes sense, look back at the explanation of resistance and heating on page 159. If the battery voltage is increased, the electric field in the wires gets bigger. So the free charges (the electrons) move twice as fast. When a charge collides with an atom in the wire, twice as much energy is transferred in the collision. These collisions also happen more often, simply because the charges are moving along faster. So when the voltage is increased, collisions between electrons and the lattice of atoms are *both* harder *and* more frequent.

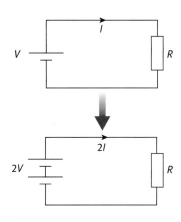

Doubling the battery voltage makes the current double. So the power ($P = IV$) is four times as big.

Questions

1 In these two circuits, resistor R_1 has a large resistance and resistor R_2 has a small resistance. If each circuit is switched on for a while, which resistor will get hotter? Explain your answers.

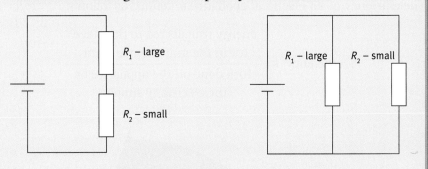

2 In circuit A, a battery is connected to a resistor with a small resistance. In circuit B, the resistor has a large resistance. The two batteries are identical. Which will go 'flat' first? Explain your answer.

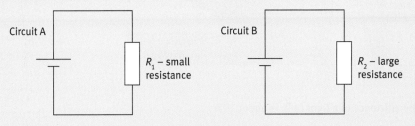

The power of the electric motor in this tube train is much greater than the power of the strip light above the platform. Both the voltage and the current are bigger.

Key words

power

Find out about:

▶ how to calculate the energy transferred by a domestic appliance

▶ what is meant by 'efficiency'

▶ how to calculate the correct fuse value for an appliance

Every electrical appliance has a power rating (in W) marked on it. This tells you how much work is done by the electricity supply every second it is on.

The compact fluorescent (energy-saving) bulb on the left is more efficient than the filament lamp on the right. The electricity supply has to do less work to produce the same amount of light output per second.

G Domestic appliances

Electricity bills are based on the amount of 'electrical energy' used. This is the amount of work done by the electricity supply on all the appliances we run. The work done on an appliance depends on

▶ its power rating

▶ the time it is on for

You could calculate this in joules, but the result would usually be a very large number. So, for domestic appliances, it is more convenient to use the **kilowatt-hour** as the unit of energy:

$$\begin{array}{ccc} \text{energy transferred} & = & \text{power rating} \quad \times \quad \text{time} \\ \text{when device is on} & & \end{array}$$

$$\begin{array}{ccc} \text{(joule, J)} & \text{(watt, W)} & \text{(second, S)} \\ \text{(kilowatt-hour, kW h)} & \text{(kilowatt, kW)} & \text{(hour, h)} \end{array}$$

On an electricity bill, '1 unit' means 1 kilowatt-hour. The electricity meter in your house measures the number of kilowatt-hours of electrical energy you buy.

Efficiency

In all electrical appliances, some of the energy transferred does not end up where it is wanted, or in the form it is wanted. For example, in a filament light bulb, less than 10% of the work done on the filament is carried away as light. The rest goes into heating the bulb and its surroundings. The **efficiency** of an electrical appliance is defined as follows:

$$\% \text{ efficiency} = \frac{\begin{array}{c}\text{energy transferred to the place} \\ \text{(or in the manner) we want}\end{array}}{\begin{array}{c}\text{work done on the appliance by} \\ \text{the electricity supply}\end{array}}$$

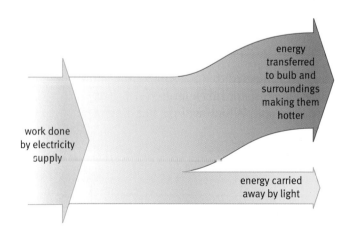

The efficiency of a light bulb is about 10%.

Working out the current

If you know the power rating of a mains appliance, you can easily work out the current through it when it is switched on, using the equation.

$$\text{power} = \text{current} \times \text{voltage}$$
$$\text{(watt, W)} \quad \text{(ampere, A)} \quad \text{(volt, V)}$$

For all UK mains appliances, the operating voltage is 230 V. So

$$\text{power (in W)} = \text{current (in A)} \times 230\,\text{V}$$

Divide both sides by 230 V to find the current:

$$\text{current (in A)} = \frac{\text{power (in W)}}{230\,\text{V}}$$

This is important to know when you have to choose the right fuse for a plug. A fuse is a short piece of wire made of a metal that melts at a low temperature. An electric current through it makes it heat up. Its length and thickness are chosen so that it melts if the current goes above the value marked on the fuse. So it is, the 'weakest link' in the circuit. It will melt first if the current for any reason gets bigger than it should be. Fuses for mains plugs normally come in two values: 3 A and 13 A. The electricity companies recommend using a 3 A fuse for any appliance with a power rating below 690 W, and a 13 A fuse for those with a higher power rating.

> **Key words**
>
> kilowatt-hour
> efficiency

Questions

1 What is the power of a mains appliance that needs a current of 3 A to make it run? Use your answer to explain the advice of the electricity companies on choosing fuse values.

2 Look carefully at each of the tasks in the table on the right which use electricity, and put them in the order you think they would come – from the cheapest to the most expensive. Then calculate the number of kilowatt-hours which each involves and see if your estimate of the cost was correct.

Task	Appliance used	Power rating (W)	Time for which it is on
watch television for the evening	television	300	5 hours
dry your hair	hairdryer	700	5 minutes
make a pot of tea	electric kettle	2000	4 minutes
write a homework assignment	computer	250	2 hours
keep a front door light on overnight	light bulb	100	10 hours
listen to a football match on the radio	radio	10	2 hours
heat your bedroom while you do your homework	electric fan heater	1500	2 hours
wash a load of dirty clothes	washing machine	1850	$1\frac{1}{2}$ hours

3 A 20 W energy-saving light bulb costs £2.99. It gives the same amount of light as a 100 W filament light bulb, costing 45p. The lifetime of the energy-saving bulb is 5000 hours. The filament bulb has a lifetime of 1000 hours. If 1 unit (1 kilowatt-hour) of electrical energy costs you 5p, how much do you pay for 5000 hours of lighting with each bulb? Which is the better buy?

Find out about

▶ how a magnet moving near a coil can generate an electric current
▶ the factors that determine the size of this current
▶ how this is used to generate electricity on the large scale

H An electricity supply

Nowadays most of us in Britain take a mains electricity supply for granted. But in fact it was only in May 2003 that Cym Brefi in mid-Wales became the last village in Britain to get a mains electricity supply.

Electricity reaches last village in Britain

Of course, not having a mains electricity supply does not mean you cannot use electrical appliances. Many can be run from batteries. But this works only for relatively low-power devices. For others, you might use a diesel-powered **generator**.

This is how the inhabitants of Cym Brefi ran their washing machines and vacuum cleaners before they got mains electricity. But generators are noisy, and each 'unit of electricity' is much more expensive than from the mains. So they can only be run for a short time.

Generating electricity

Generators work on the principle of **electromagnetic induction**. This phenomenon, which does so much to make our lives comfortable and convenient, was discovered in the 1830s by Michael Faraday.

About 10 years earlier, a Danish physicist, Hans Christian Oersted, had shown that an electric current in a wire produces a magnetic field in the region around it. Faraday wondered if he could do this in reverse: use a magnetic field to produce an electric current. After several years of experimentation, using homemade coils of wire and magnets, he succeeded.

One way to generate a current is to move a magnet into, or out of, a coil. The movement of the magnet causes an induced voltage across the ends of the coil. 'Induced' means that it is caused by something else – in this case, the movement of the magnet. The coil, for a brief time, is like a small battery. If the coil is part of a closed circuit, this induced voltage makes a current flow.

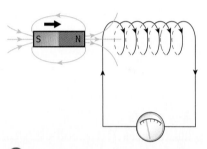

1 While the bar magnet is moving into the coil, there is a small reading on the sensitive ammeter.

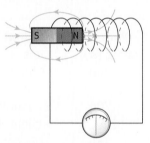

2 There is no current while the magnet is stationary inside the coil.

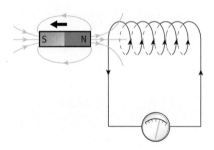

3 While the magnet is being removed from the coil, there is again a small current, but now in the opposite direction.

Moving a magnet into, or out of, a coil generates a current.

The size of the current can be increased by

- moving the magnet in and out more quickly

- using a stronger magnet

- using a coil with more turns

Now imagine what would happen if the magnet were rotated near the end of the coil. The magnetic field around the coil would be constantly reversing direction. This changing magnetic field induces a current in the coil, first in one direction, then the other. This is called an **alternating current** (a.c.). For many applications, this works just as well as a **direct current** (d.c.). A direct current is one which is always in the same direction.

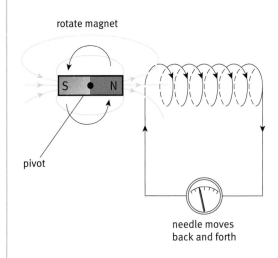

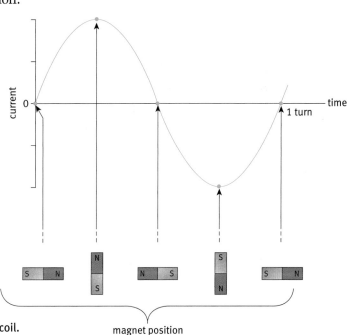

If the magnet is rotated, it will induce an alternating current in the coil.

The size of the alternating voltage (and current) produced by a generator of this sort can be increased by:

- using a stronger rotating magnet or electromagnet

- rotating the magnet or electromagnet faster (though this also affects the frequency of the a.c. produced)

- using a fixed coil with more turns

- putting an iron core inside the fixed coil (this makes the magnetic field a lot bigger – as much as 1000 times)

In a real generator, an electromagnet is rotated inside a fixed coil. As it spins, a.c. is generated in the coil. In power stations in the UK, the rate of turning is set at 50 cycles per second. The generator is turned by a turbine, which is driven by steam. The steam is produced by burning gas, oil, or coal, or by the heating effect of a nuclear reaction.

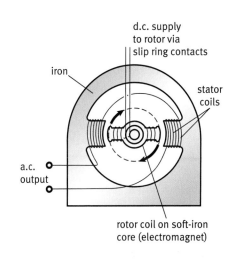

A simplified diagram of an a.c. generator

171

Distributing electricity

Transformers

An electric current can be generated by moving a magnet into (or out of) a coil of wire (see page 170). The moving magnet could be replaced by an electromagnet. If a coil is wound round an iron core, it becomes quite a strong magnet when a current flows through it.

(see page 170)

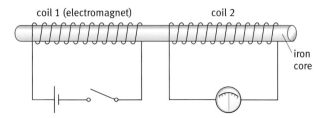

When the current in the electromagnet (coil 1) is switched on, this has the same effect as plunging a bar magnet into coil 2. So a current is generated in coil 2, whilst the current in coil 1 is changing. This arrangement of two coils on the same iron core is called a **transformer**. Changing the current in the primary coil induces a voltage across the secondary coil. If the current in the primary is a.c., it is changing all the time. So an alternating voltage is induced across the secondary coil.

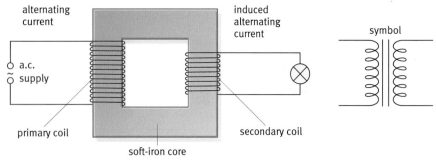

The transformer. When the current in the primary coil is changing, a voltage is induced across the secondary coil. This makes a current flow round the right-hand circuit. Notice that there is no direct electrical connection between the coils of a transformer. The only connection is through the magnetic field.

The behaviour of a transformer depends on the number of turns of wire on the two coils.

The equation that links the two is:

$$\frac{\text{voltage across secondary coil } (V_S)}{\text{voltage across primary coil } (V_P)} = \frac{\text{number of turns on secondary coil } (N_S)}{\text{number of turns on primary coil } (N_P)}$$

If there are more turns on the secondary coil, then the induced voltage across this coil is bigger than the applied voltage across the primary coil. However, this is not something for nothing! The current in the secondary will be correspondingly less, so that the power available from the secondary is no greater than the power supplied to the primary (remember: power = IV).

Find out about:

▶ how transformers are used to alter the voltage of a supply

▶ the main components of the National Grid

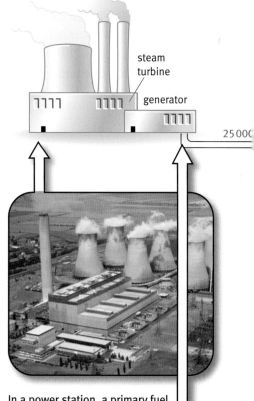

In a power station, a primary fuel is used to generate electricity. The commonest fuel is gas, but coal, nuclear fuel, water in high dams, and wind are also used.

The heart of a power station is a turbine. The primary fuel is used to drive this, and as it turns it makes the coil of a generator rotate.

The National Grid

Transformers play an important role in the National Grid system for distributing electricity. All the electricity power stations in Britain are connected into the National Grid, which is used to distribute electricity to all the places where we want to use it. The Grid connects every power socket in your home back to the power station. It does this by means of a long chain of wires and magnetic fields in transformers.

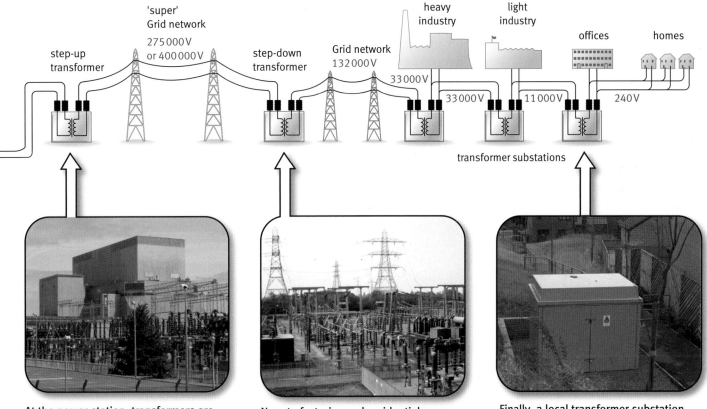

At the power station, transformers are used to raise the voltage to a very high value (sometimes as much as 400 000 V). This means that the current in the pylon lines is small. So relatively little energy is wasted heating the pylon cables themselves.

Near to factories and residential areas, transformer substations reduce the voltage to a lower level, around 33 000 V.

Finally, a local transformer substation reduces the voltage to 230 V. This is the voltage at which electricity is supplied to homes in the UK. There is likely to be one of these transformer substations close to where you live.

Questions

1 A transformer has 100 turns on its primary coil, and 25 turns on its secondary coil. A 12 V a.c. supply is connected to the primary coil. What will be the voltage across the secondary coil?

2 If you had a 6 V a.c. supply and wanted to use it to operate a 12 V bulb, explain how you could make a simple transformer to enable you to do this.

3 In the National Grid, transformers are used to 'step-up' the voltage from 25 000 V to 400 000 V. What gets smaller as a result (and stops us getting something for nothing)?

Summary

In this module you have met some of the basic ideas that scientists use to explain electrical phenomena: charge, current, voltage, resistance, and potential difference. The module has introduced you to some models that are useful for explaining and predicting the behaviour of electric circuits.

Electric charge

▶ Electric charge is a fundamental property of matter.

▶ Charge cannot be created or destroyed. But positive and negative charges can be separated, and moved from one object to another, for example by rubbing.

Electric current

▶ A working electric circuit always consists of a closed loop (or loops) of conducting material, between the positive and negative terminals of a battery.

▶ An electric current is a flow of charges, which are already present in the materials of the circuit. The battery makes the charges move round the circuit.

▶ Current is not used up as it goes round – but it does work on the components it passes through, and so transfers energy from the battery to the other components.

Current, voltage, and resistance

▶ The voltage of a battery is a measure of the strength of its 'push' on the charges. The bigger the voltage, the bigger the current.

▶ The components in a circuit resist the flow of charge. The bigger the resistance, the smaller the current.

▶ Together, the battery voltage and the circuit resistance determine the current in the circuit.

Resistors in series and parallel

▶ The total resistance of resistors in series is their sum.

▶ The total resistance of resistors in parallel is less than that of any single resistor – as the group provides more loops for charges to flow round. H

Potential difference

▶ It is useful to think of a battery as raising charges to a higher level (giving them more potential energy). They then lose this potential energy as they go round the circuit.

▶ A voltmeter measures the potential difference (p.d.) between the two points it is connected to.

▶ Resistors in parallel have the same p.d. across each of them.

▶ The p.d. across resistors in series is proportional to their resistance.

Electrical power

▶ The power (energy per second) transferred by an electric circuit is equal to 'current × voltage'.

Electromagnetic induction

▶ A potential difference (p.d.) is induced across the ends of a wire or coil placed in a changing magnetic field.

▶ If this wire or coil is part of a circuit, there is an induced electric current in the circuit.

▶ This phenomenon is called electromagnetic induction. It is the basis of the electrical generator, and the transformer – both of which are key components of the mains electricity supply system (the National Grid).

Questions

1   Look at the three electric circuit models on page 154. Copy and complete the following table:

Model	What corresponds to the battery?	What corresponds to electric current?	What corresponds to the resistors or lamps in the circuit?
'peas in a pipe'			
'chain on a bike'			
'moving rope'			

2   In a simple single-loop electric circuit, the current is the same everywhere. It is not used up. How does each of the models above help to account for this?

3   Imagine a simple electric circuit consisting of a battery and a bulb. For each of the following statements, say if it is true or false (and explain why):

   a Before the battery is connected, there are no electric charges in the wire. When the circuit is switched on, electric charges flow out of the battery into the wire.

   b Collisions between the moving charges and fixed atoms in bulb filament make it heat up and light.

   c Electric charges are used up in the bulb to make it light.

4   In shops, you can buy batteries labelled 1.5 V, 4.5 V, 6 V, or 9 V. But you cannot buy batteries labelled 1.5 A, 4.5 A, 6 A or 9 A. Explain why not.

5   You are given four 4 Ω resistors. Draw diagrams to show how you could connect all four together to make a resistance of:

   a 16 Ω

   b 5 Ω

   c 2 Ω (there are two ways of doing this)

   d 1 Ω

6   A family's electric shower has an electrical power of 6 kW. In a typical week, it is used for twelve showers, each lasting ten minutes.

   a Use the equation
      energy (joules) = power (watts) × time (seconds)
      to calculate the energy (in joules) transferred by the shower in a typical week.

   b Use the equation energy (kilowatt hours) = power (kilowatts) × time (hours) to calculate the energy (in kWh) transferred by the shower in a typical week.

   c The electricity company charges 8p per kWh. Calculate how much the twelve showers cost.

7   By my bed I have a 60 W spotlight, which I use for reading. It is only about 10% efficient. My tortoise has an identical 60 W spotlight which he basks under. It is about 90% efficient. Explain why both of these statements can be true.

8   Copy and complete these sentences:

   When a magnet is moved into a coil of wire, a voltage is _____ in the coil. The voltage is produced only when the magnet is _____ . This is used in an a.c. generator, which has a _____ rotating near a coil (see diagram on page 171). To increase the size of the induced voltage, you could use a _____ magnet, have more _____ on the coil, turn the coil _____ , or put a core of _____ inside it.

   The current in the external circuit constantly changes direction, so it is called _____ current (_____). This is differerent from the current from a battery, which always goes in one direction and is called _____ current (_____).

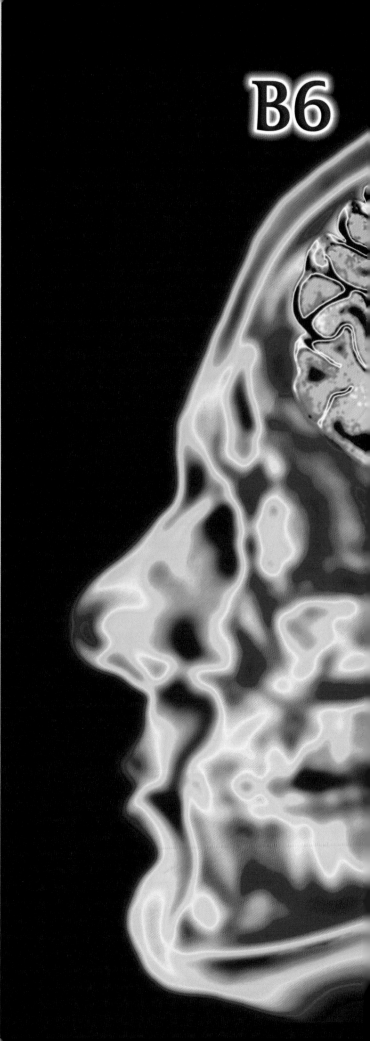

Why study the brain and mind?

The human brain allows our species to survive on Earth. It gives us advantages of intelligence and sophisticated behaviour. Without our complex brains our species may easily have died out thousands of years ago.

The science

Animals respond to stimuli in order to survive. The central nervous system, the brain and spinal cord, coordinates millions of electrical impulses every second. These impulses determine how we think, feel, and react – our behaviour. Some drugs can affect our behaviour by interfering with the way nerve cells carry impulses.

The structure of our brains allows human beings to learn from experience and recall large amounts of information from our memory stores. Scientists have come up with some models for memory, but so far none can really fully explain how it works.

Biology in action

Research into the brain is helping to develop new treatments for some diseases. Knowing more about how we learn could help many people, for example children with learning difficulties, and adults recovering from brain damage.

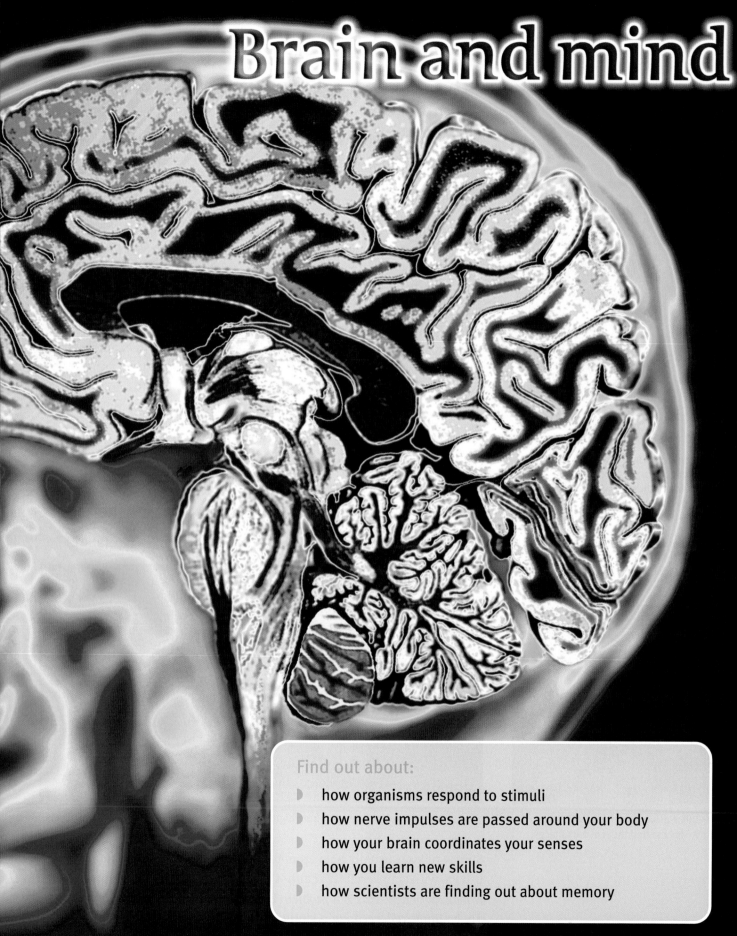

Brain and mind

Find out about:

- how organisms respond to stimuli
- how nerve impulses are passed around your body
- how your brain coordinates your senses
- how you learn new skills
- how scientists are finding out about memory

Find out about:

▶ what behaviour is
▶ how simple behaviour helps animals survive

A What is behaviour?

Imagine you are sitting outside. The temperature drops, and you get cold. You start to shiver.

Shivering is a **response** to the change in temperature. A change in your environment, like a drop in temperature, is called a **stimulus**. Eating is a response to the stimulus of hunger. Scratching is a response to an itch. Shivering, eating, and scratching are all examples of **behaviour**.

You can think of behaviour as anything an animal does. The way an animal responds to changes in its surroundings is important for its survival.

Simple behaviour

Simple animals always respond to a stimulus in the same way. For example, woodlice always move away from light. This is an example of a **simple reflex** response. Reflexes are always **involuntary** – they are automatic. Reflexes are important because they increase the animal's chance of survival. The photographs in this section all show reflexes.

Why are simple reflexes important?

Simple reflex behaviour helps an animal to:

▶ find food, shelter, or a mate

▶ escape from predators

▶ avoid harmful environments, for example extreme temperatures

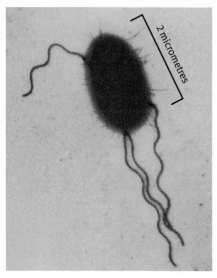

Escherichia coli bacteria are found in the lower gut of warm-blooded animals. They detect the highest concentration of food and move towards it.

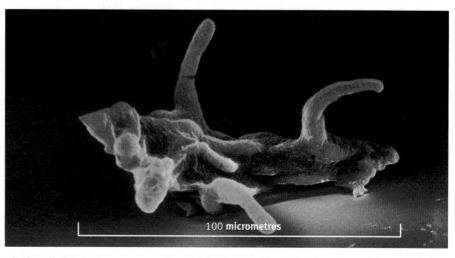

Single-celled *Amoeba* move away from high concentrations of salt, strong acids, and alkalis.

Woodlice move away from light, so you are most likely to find them in dark places.

Simple reflexes usually help animals to survive. But animals that only behave with simple reflexes cannot easily change their behaviour, or learn from experience. This is a problem if conditions around them change. Their simple reflexes may no longer be helpful for survival.

When a giant octopus sees a predator, it rapidly contracts its body muscles. This squirts out a jet of water to push the octopus away from danger. The octopus may also release a dark chemical (often called 'ink') which hides its escape.

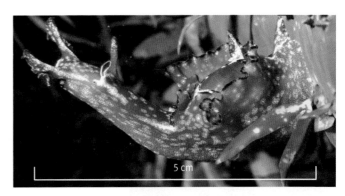

When the tail of the sea hare *Aplysia* is pinched, the muscle contracts quickly and strongly. This reflex helps the animal escape from the spiny lobster that preys on it.

Earthworms have some of the fastest reflexes in the animal kingdom. A sharp tap from a beak on its head end is detected in the body wall. Rapid contraction of the worm's muscles pulls it back into its burrow. But this time the bird was too quick.

Have you ever tried to swat a fly? It has a very fast response to any movement that its sensitive eyes detect.

Complex behaviour – a better chance of survival

Complex animals, like mammals, birds, and fish, have simple reflex responses. But a lot of their behaviour is far more complicated. It includes reflex responses that have been altered by experience. Also, much of their behaviour is not involuntary – they make conscious decisions. For example, if it gets very cold, you do not just rely on your reflexes to keep you warm; you decide to put on extra clothes.

Because complex animals can change their behaviour when environmental conditions change, they are more likely to survive.

> **Key words**
>
> response
> stimulus
> behaviour
> simple reflex
> involuntary

> **Questions**
>
> 1 Write a sentence to explain each key word on this page.
>
> 2 Describe an action you did today that:
>   - you did not have to think about and you have never had to learn to do
>   - you have learned to do but you can now do without thinking
>   - you had to think about while you were doing it
>
> 3 Which of the actions you described is an example of conscious behaviour, and which is most likely to be a simple reflex response?

Find out about:

▶ reflexes in newborn babies
▶ simple reflexes that help you to survive

B Simple reflexes in humans

Behaviour in humans and other mammals is usually very complex. But simple reflexes are still important for survival. For example:

▶ When an object touches the back of your throat, you gag to avoid swallowing it. This is the gag reflex.

▶ When a bright light shines in your eye, your pupil becomes smaller. This **pupil reflex** stops bright light from damaging the sensitive cells at the back of your eye.

Newborn reflexes

When a baby is born, the nurse checks for a set of **newborn reflexes**. Many of these reflexes are only present for a short time after birth. They are gradually replaced by behaviours learned from experience. In a few cases these reflexes are missing at birth, or they are still present when they should have disappeared. This may mean that the baby's nervous system is not developing properly.

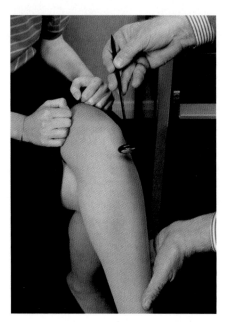

The knee jerk reflex causes your thigh muscle to contract, so your lower leg moves upwards. Doctors may test this and other reflexes when you have a health check. Try standing still with your eyes closed. You will notice this reflex helping you to balance.

Stepping. If you hold a baby under his arms, support his head, and allow his feet to touch a flat surface, he will appear to take steps and walk. This reflex usually disappears by 2–3 months after birth. It then reappears as he learns to walk at around 10–15 months.

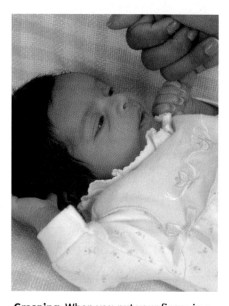

Grasping. When you put your finger in a baby's open palm, the baby grips the finger. When you pull away, the grip gets stronger. This reflex usually disappears by 5–6 months. If you stroke the underneath of a baby's foot, its toes and foot will curl. This reflex usually disappears by 9–12 months.

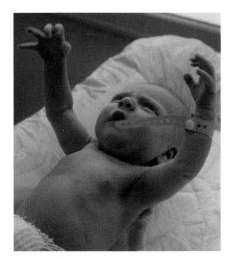

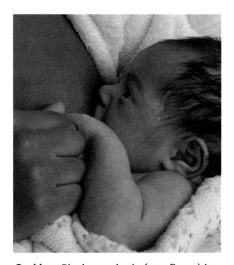

Startle. This is also called the Moro reflex, named in 1687 after the Italian scientist Artur Moro. It usually happens when a baby hears a loud noise or is moved quickly. The response includes spreading the arms and legs out and extending the neck. The baby then quickly brings her arms back together and cries. This reflex usually goes by 3–6 months.

Sucking. Placing a nipple (or a finger) in a baby's mouth causes the sucking reflex. It is slowly replaced by voluntary sucking at around 2 months.
Rooting. Stroking a baby's cheek makes her turn towards you, looking for food. This reflex helps the baby find the nipple when she is breast feeding. The rooting reflex is gone by about 4 months.

Swimming. If you put a baby under six months of age in water, he moves his arms and legs while holding his breath.

Sudden infant death syndrome (SIDS)

Sudden Infant Death Syndrome, or cot death, is tragic and unsolved. In the UK about 7 babies a week die from SIDS. This is 0.7 deaths for every 1000 live births.

It is likely that there are many different causes of cot death. Some people think that it could be because a baby's simple reflexes have not matured properly. This is how doctors think this may happen:

▸ When a fetus detects that oxygen in its blood is low, its reflex response makes it move around less. This makes sense because the less it moves around, the less oxygen it will use up in cell respiration.

▸ This response changes as the baby matures. When an older baby or child's airways are covered, for example, by a duvet, the baby moves more. He turns his head from side to side. He also pushes the obstruction away. So now the response to low oxygen is more activity, not less.

▸ If the newborn baby has not grown out of the fetal reflex, he may lie still if his bedcovers cover his airways. He is more likely to suffocate.

Doctors now advise mothers to put babies onto their backs to sleep, and not to use soft bedding like duvets. This way their faces are less likely to become covered.

Key words

pupil reflex
newborn reflexes

Questions

1 Describe two reflexes in:

  a adult humans

  b newborn babies

2 How do you think the startle reflex helps a baby to survive?

3 Why are premature babies more at risk from SIDS than babies born at the correct time?

Find out about:

▶ different parts of your nervous system
▶ how reflexes are controlled

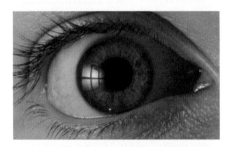

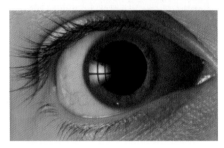

Muscles in the iris cause the pupil to change size depending on the brightness of light entering the eye. (The pupil size controls the amount of light intensity that is the stimulus.)

Key words

nervous system
nerve impulses
reflex arc
receptor
sensory neuron
central nervous system
coordinates
motor neuron
effector
peripheral nervous system
neurons
axon
fatty sheath

c Your nervous system

Walk out from a dark cinema on a bright afternoon and your pupils will become smaller. This pupil reflex prevents bright light damaging your eye. Like all reflexes, this behaviour is coordinated by your **nervous system**.

Cells in your nervous system carry **nerve impulses**. These nerve impulses allow the different parts of the nervous system to communicate with each other.

The reflex arc

In a simple reflex, impulses are passed from one part of the nervous system to the next in a pathway called a **reflex arc**. The diagram below shows this pathway for the pupil reflex:

▶ The stimulus is detected by a **receptor** cell.

▶ Nerve impulses are carried along a **sensory neuron** to your **central nervous system**.

▶ The central nervous system is made up of your brain and spinal cord. It **coordinates** your body's responses.

▶ Nerve impulses are carried along a **motor neuron** to an **effector**.

▶ The effector carries out the response to the stimulus.

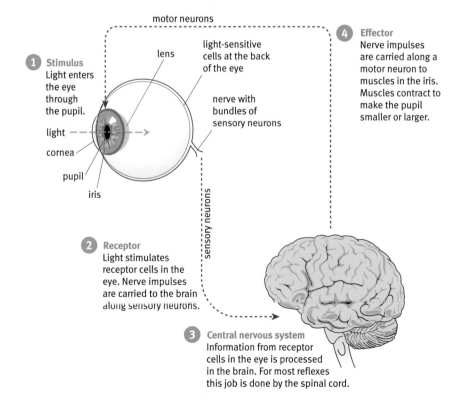

An example of a reflex arc

Peripheral nervous system

Many nerves link your brain and spinal cord to every other part of your body. These nerves make up the **peripheral nervous system**.

Nerves and neurons

Nerves are bundles of specialized cells called neurons. Like most body cells, **neurons** have a nucleus, a cell membrane, and cytoplasm. They are different from other cells because the cytoplasm is shaped into a very long thin extension. This is called the **axon**, and it is how neurons connect different parts of the body.

Axons carry electrical nerve impulses. Like wiring in an electrical circuit, the axons must be insulated from each other. The insulation for an axon is a **fatty sheath** wrapped around the outside of the cell. The fatty sheath increases the speed that impulses move along the axon.

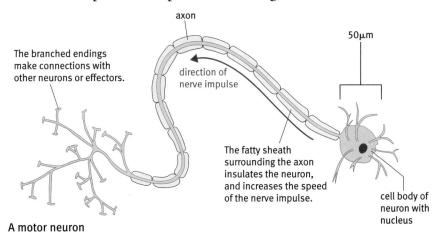

axon

50μm

The branched endings make connections with other neurons or effectors.

direction of nerve impulse

The fatty sheath surrounding the axon insulates the neuron, and increases the speed of the nerve impulse.

cell body of neuron with nucleus

A motor neuron

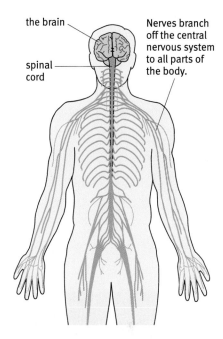

the brain

spinal cord

Nerves branch off the central nervous system to all parts of the body.

The peripheral nervous system links the brain and spinal cord with the rest of the body.

Central nervous system

Your central nervous system (CNS) coordinates all the information it receives from your receptors. Information about a stimulus goes to either your brain or your spinal cord. In a reflex arc the CNS directly links the incoming information from receptors with the effectors that will carry out the necessary response.

Questions

1 Describe the difference between:
  ▶ the job of a sensory neuron and a motor neuron
  ▶ an axon and a neuron
  ▶ the central nervous system and the peripheral nervous system

2 Draw a labelled diagram to show a reflex arc for the newborn grasp reflex shown on page 180. This reflex is coordinated by the spinal cord.

Receptors

You can only respond to a change if you can detect it. Receptors inside and outside your body detect **stimuli**, or changes in the environment.

Some animals have receptors that are triggered by stimuli which humans cannot detect. For example, falcons can detect an object 10 cm across when they are 1.5 km away. The rattlesnake's heat-sensitive receptors can detect a warm mouse 40 cm away. The receptors in your eyes and skin could not detect these stimuli.

Some sharks can detect when the concentration of food in the water is only 0.1 parts per billion. This concentration certainly would not trigger your taste or smell receptors. Sharks also have receptors that are stimulated by the tiny electrical currents given off by their prey.

A shark's sensitive receptors help it to locate its prey.

The falcon's eyes can spot the slightest movement on the ground.

Body heat gave the position of this rattlesnake's latest meal.

Detecting stimuli inside and out

You can detect many different stimuli, for example sound, texture, smell, temperature, and light. Different types of receptors each detect a different type of stimulus. Receptors on the outside of your body monitor the external environment. Others monitor changes inside your body, for example core temperature and blood sugar levels.

Sense organs

Some receptors are made up of single cells, for example pain receptors in your skin. Other receptor cells are grouped together as part of a complex sense organ, for example your eye. Vision is very important in humans and most other mammals. Light entering our eyes helps us humans produce a three-dimensional picture of our surroundings. This gives us information about objects such as their shape, movement, and colour.

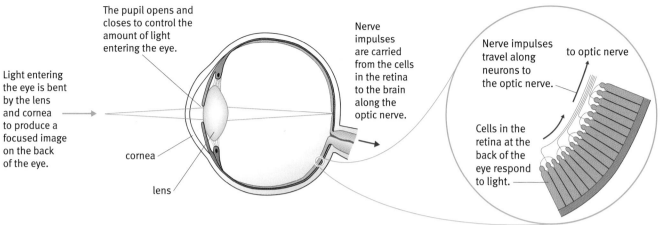

The pupil opens and closes to control the amount of light entering the eye.

Light entering the eye is bent by the lens and cornea to produce a focused image on the back of the eye.

cornea

lens

Nerve impulses are carried from the cells in the retina to the brain along the optic nerve.

Nerve impulses travel along neurons to the optic nerve.

to optic nerve

Cells in the retina at the back of the eye respond to light.

Light is focused by the cornea and lens onto light-sensitive cells at the back of the eye. These cells are receptors. They trigger nerve impulses to the brain.

Effectors

The body's responses to stimuli are carried out by effector organs. Effectors are either `glands` or `muscles`. When impulses arrive at these effectors they cause:

▶ glands to release chemicals, for example hormones, enzymes, or sweat

▶ muscles to contract and move part of the body

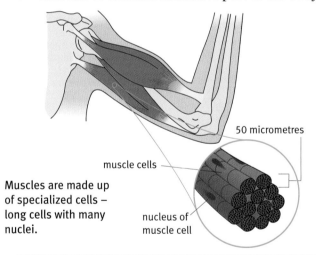

Muscles are made up of specialized cells – long cells with many nuclei.

muscle cells

nucleus of muscle cell

50 micrometres

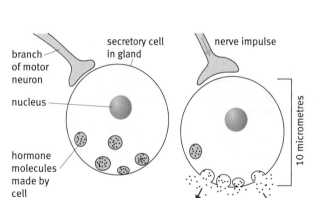

branch of motor neuron

nucleus

hormone molecules made by cell

secretory cell in gland

nerve impulse

to the blood stream

10 micrometres

Hormone molecules are released by secretory cells when the gland is stimulated by a nerve impulse.

Questions

3 What kind of stimulus can a shark detect that we cannot?

4 Name the two different types of effectors and say what they do

5 Which receptors would you use in order to thread your trainers with new laces?

6 Which effectors are you using when you:

▶ text a friend?

▶ cry?

▶ run a race?

7 Some people suffer from a disease where tiny clusters of light receptor cells in different parts of the eye become damaged. How would this affect what the person sees?

Conscious control of reflexes

Reflexes are involuntary actions. Most of them are coordinated by your spinal cord. Even those that are coordinated by your brain, for example your pupil reflex, are involuntary. You do not think about these responses – your brain does not have to make a decision. They happen automatically, because they are designed to help you survive.

But sometimes a reflex may not be what you want to happen. Some reflexes can be modified by conscious control. Imagine picking up a hot plate. Your pain reflex makes you drop it. But if your dinner were on the plate, you can overcome this reflex and hold onto the plate until you were able to put it down safely. The conscious control of your brain overcomes the reflex response. The diagram below explains how this happens.

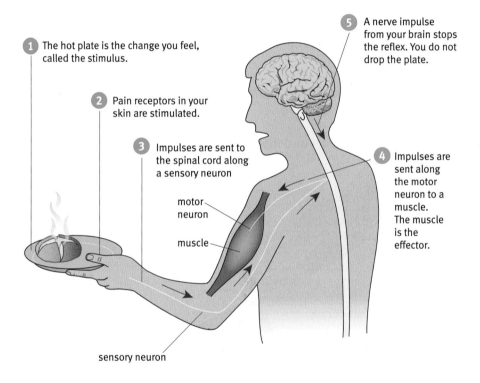

1 The hot plate is the change you feel, called the stimulus.

2 Pain receptors in your skin are stimulated.

3 Impulses are sent to the spinal cord along a sensory neuron

5 A nerve impulse from your brain stops the reflex. You do not drop the plate.

4 Impulses are sent along the motor neuron to a muscle. The muscle is the effector.

motor neuron

muscle

sensory neuron

Overcoming a reflex by conscious control. Some reflexes can be modified by conscious control. If you pick up a hot plate, your reflex response is to drop it. But a second nerve impulse travels up your spine to your brain, then back down to cause a muscle movement stopping this reflex. You can put the plate down quickly without dropping it because the conscious control from your brain overcomes the reflex.

Questions

8 List two reflexes you can overcome, and two that you can not.

9 A tiny baby urinates whenever its bladder is full. Draw a labelled diagram to show how nerve impulses from the brain overcome this reflex when he is older.

It's all in the mind – more complex behaviour

Most human reflex arcs are coordinated by the spinal cord. Only reflexes with receptors on the head are coordinated by the brain. A reflex arc only has simple connections between a sensory neuron and a motor neuron.

Connections in the brain usually involve hundreds of other neurons with different connections. Using these complex pathways, your brain can process highly complicated information, such as music, smells, and moving pictures. Different parts of the brain also store information (memory) and use it to make decisions for more complicated behaviour.

You can find out more about the brain, and complex behaviour like learning and memory, later in this module.

Hundreds of neurons interact to coordinate the responses you make when you are receiving this many stimuli.

Hundreds of complex pathways in the brain have to be used to succeed in this fast-moving sport.

This complicated behaviour involves highly complex pathways in the brains of both these animals.

A day in the life of a neuroscientist

Steven Rose is a neuroscientist – a scientist who studies how the brain works. Steven is interested in how complex behaviour is controlled.

He studies the behaviour of chicks and looks for changes in chemicals in their brains as they learn new skills.

From his work with chicks, Professor Rose found that a particular chemical, APP, is needed for learning and memory. If this chemical is changed or absent, animals cannot remember something new.

A small amount of APP can rescue the memory of animals with symptoms similar to Alzheimer's disease (loss of memory). This research is leading to a new approach to drug development for Alzheimer's.

Quick responses to stimuli are also essential in real life. They help you survive by avoiding danger.

Curare is a very powerful toxin. It is used on the tips of blowpipe darts.

D Synapses

Think about playing a fast sport or computer game. You need very quick reactions to win. Nerve impulses give you fast reactions because they travel along axons at 400 metres per second.

Mind the gap

Neurons do not touch each other. So when nerve impulses pass from one neuron to the next, they have to cross tiny gaps. These gaps are called **synapses**. Some drugs and poisons (toxins) interfere with nerve impulses crossing a synapse. This is how they afffect the human body.

How do nerve impulses cross a synapse?

Nerve impulses cannot jump across a synapse. Instead, chemicals are used to pass an impulse from one neuron to the next. The diagram below explains how this works.

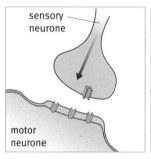
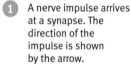
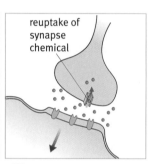

1 A nerve impulse arrives at a synapse. The direction of the impulse is shown by the arrow.

2 A chemical is released from the sensory neuron. It diffuses across the synapse.
The molecule is the correct shape to fit into **receptor molecules** on the membrane of the motor neuron.

3 A nerve impulse is stimulated in the motor neuron. The chemical is absorbed back into the sensory neuron to be used again.

How a synapse works

Do synapses slow down nerve impulses?

The gap at a synapse is only about 20 **nanometres** (nm) wide. The synapse chemical travels across this gap in a very short time. Synapses do slow down nerve impulses to about 15 metres per second. A nerve impulse still travels from one part of your body to another at an incredible speed.

Being human – just chemicals in your brain?

The way we think, feel, and behave does involve a series of chemicals moving across synapses between neurons, but there is more to behaving like a human than chemicals in your brain. These processes are very complicated. Scientists researching how your nervous system works are only just beginning to understand the brain.

Serotonin

Serotonin is a chemical released at one type of synapse in the brain. When serotonin is released, you get feelings of pleasure. Pleasure is an important response for survival. For example, eating nice-tasting food gives you a feeling of pleasure. So you are more likely to repeat eating, which is essential for survival.

Lack of serotonin in the brain is linked to depression. Depression is a very serious illness. Approximately X million people a year in the UK are diagnosed with depression. They feel very unhappy for many days on end and often find it difficult to manage normal everyday things like working, studying, or looking after their family.

How do some drugs affect the brain?

Prozac is the name of an antidepressant drug. These drugs can be helpful for treating people with depression. Prozac causes serotonin concentration to build up in synapses in the brain. So a person may feel less unhappy. The diagram explains how Prozac works. Like all drugs, Prozac can have unwanted effects. A doctor will consider all the factors very carefully before prescribing treatment.

Ecstasy

Ecstasy is the common name for the drug MDMA. Ecstasy works in a similar way to Prozac. People who have taken Ecstasy say that it can give them feelings of happiness and being very close to other people. Studies on monkeys suggest that long-term use of Ecstasy may destroy the synapses in the pleasure pathways of the brain. Permanent anxiety and depression might result, along with poor attention span and memory. For some people the harmful effects of Ecstasy are more immediate. Ecstasy interferes with the body's temperature control systems. It also slows down production of the hormone ADH in the brain. These effects can be fatal. You can read more about ADH in Module B4 *Homeostasis*.

Prozac molecule blocking reuptake of serotonin from a synapse

Feelings of depression can be caused by too little serotonin in the brain. Prozac works by blocking reuptake of serotonin.

Questions

1 Write down a sentence to describe a synapse.

2 Draw a flow diagram to describe what happens when a nerve impulse arrives at a synapse.

3 Explain how the release of serotonin in the brain helps us survive.

4 Some drugs (like curare) block the receptors on the motor neuron at a synapse. Explain how this would affect muscles that the motor neuron is linked to.

5 Explain how Prozac may help a person who has depression.

Key words

synapses
receptor molecules
nanometres
serotonin
Prozac
Ecstasy

Find out about:

▶ the structure of your brain
▶ how scientists learn about the brain

E The brain

Think of some things you did today. Getting up, deciding what to have for breakfast, travelling to school, talking and listening to friends. All of this complex behaviour has been controlled by your brain. But exactly how it happens is still being researched. Scientists who study the brain are called **neuroscientists**. Neuroscience is a fairly 'new' science. This means that scientists have only recently started to investigate how the brain works.

Simple animals

Neurons carry electrical impulses around your body. Quite simple animals have a larger mass of neurons at one end of their body – the head end. This end reaches new places first, as the animal moves. These neurons act as a simple brain. It processes information coming from the receptors on their head end.

Looking at how the brains of simple animals work can help scientists begin to understand more complicated brains.

Complex animals

More complex behaviour like yours needs a much larger brain. So your brain is made of billions of neurons. It also has many areas, each carrying out one or more specific functions all in the same organ. Your complex brain allows you to learn from experience, for example how to behave with other people.

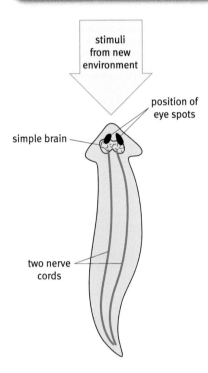

Sense organs on the flatworm head detect light and chemical stimuli. A simple brain processes the response.

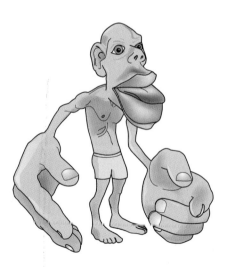

This is a diagram called the 'sensory homunculus'. Each body part is drawn so that its size represents the surface area of the sensory cortex that receives nerve impulses from it.

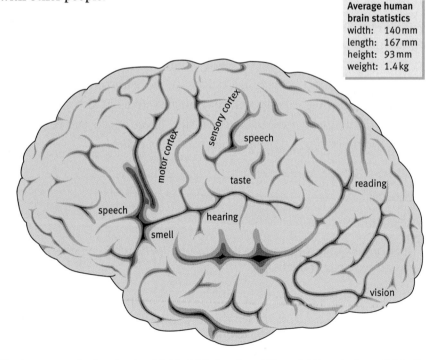

Average human brain statistics
width: 140 mm
length: 167 mm
height: 93 mm
weight: 1.4 kg

The human cerebral cortex is a highly folded region. Although it is only 5 mm thick, its total area is about 0.5 m². This map of the cerebral cortex shows the regions responsible for some of its functions.

The conscious mind

When you are awake, you are aware of yourself and your surroundings. This is called **consciousness**. The part of your brain where this happens is the **cerebral cortex**. This part is also responsible for intelligence, language, and memory. Brain processes to do with thoughts and feelings happen in the cerebral cortex and are what is called your 'mind'.

The cerebral cortex is very large in humans compared with other mammals. Studying what goes on in this part of the brain helps us to understand what it means to be human.

Finding out about the brain

In the 1940s a Canadian brain surgeon, Wilder Penfield, was working with epileptic patients. Penfield carried out operations on his patients. He applied electricity to the surface of their brains in order to find problem areas. The patients were awake during the operations. There are no pain receptors in the brain so they did not feel pain.

Penfield watched for any movement the patient made as he stimulated different brain regions. From this information he was able to identify which muscles were controlled by specific regions of the motor cortex.

Injured brains

Scientists study patients whose brains are partly destroyed by injuries or diseases like strokes. These studies provide useful information about brain function.

Brain imaging

Modern imaging techniques such as magnetic resonance imaging (MRI) scans provide detailed information about brain structure and function without having to open up the skull. MRI can be used to show up which parts of the brain are most active when a patient does different tasks. These scans are called functional MRI (fMRI) scans. The active parts of the brain have a greater flow of blood.

Key words

neuroscientists
consciousness
cerebral cortex

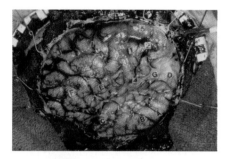

Penfield mapped the motor cortex by stimulating the exposed brain during open brain surgery. Regions of this brain have been identified and labelled in a similar way.

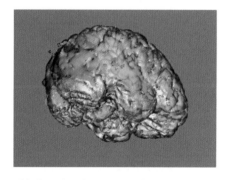

This functional MRI scan shows up areas of the brain that are active as a patient carries out a specific task. This patient was reading out loud.

Questions

1. What is your brain made up of?

2. Why is a complex brain so important for survival?

3. What four functions of the brain happen in the cerebral cortex?

4. Describe three methods that scientists have used to map the cortex.

5. Explain how the structure of your brain gives it such a large surface area.

6. Explain why it is necessary for blood flow to increase to parts of the brain that are very active.

7. Compare the diagram of the brain with the functional MRI scan. How can you tell that the person was reading aloud?

F Learned behaviour

The lion cub below is just a few weeks old. She was born with reflexes that are helping her to stay alive. But much of her behaviour for example, how to hunt, or how to get on with other lions in the pride, she will learn from her mother. This is learned behaviour.

Learned behaviour is just as important for this lion cub's survival as reflexes.

Being able to learn new behaviour by experience is very important for survival. It means that animals can change their behaviour if their environment changes.

Conditioned reflexes

In 1904 the Russian scientist Ivan Pavlov won a Nobel Prize for his study on how the digestive system works. In his research Pavlov trained a dog to expect food whenever it heard a bell ring. The diagrams on the left explain what happened.

Pavlov's experiment

Adding a stimulus that produces the same response as a reflex action is a type of learning called **conditioning**.

Pavlov's dog salivated when presented with food.

The food is the stimulus and salivation is the response.

Pavlov rang a bell while his dog was eating its food.

After a while the dog salivated when it heard the bell, even if no food was around.

The stimulus of hearing the bell became linked with food.

Learning to link a new stimulus with a reflex action allows animals to change their behaviour. This is called a **conditioned reflex.**

Conditioning aids survival

Conditioned reflexes can help animals survive. For example, bitter-tasting caterpillars are usually brightly coloured. A bird that tries to eat one learns that these bright colours mean that caterpillars will have a nasty taste. After a first experience the bird responds to the colours by leaving them alone. So this helps the caterpillars survive.

If the brightly coloured insect is also poisonous, this reflex will help the bird survive as well. If other very tasty insects have similar colours and patterns, the bird does not eat them because of this conditioned response. You might have been caught out by this too – harmless yellow-and-black striped hover flies sometimes alarm people who have been stung by a wasp.

'Warning' colours protect this caterpillar from predators.

Conditioning your pet

Open a can of soup in your kitchen. If you have a dog or a cat, this sound may get them very interested. But they are not hoping for soup! The animal's reflex response to food has been conditioned. It has learned through experience that the sound of a tin being opened may be followed by food being put into its dish.

If a cat only uses its basket when you are taking it to the vet for an injection, it may become conditioned to link the basket with a frightening experience. The cat will then always be frightened by the stimulus of the basket. It will fight to keep out of the basket, even if you are only trying to take it to a new home.

Goldfish become conditioned to expect food when they see you in the room. They swim to the front of the bowl when you appear. The goldfish are linking the stimulus of seeing you with the original stimulus – food in the water.

Questions

1 Draw a flow diagram to explain how a cat can become conditioned to expect food when it hears a bathroom shower being run.

Use the key words from this section in your answer

2 Adverts often have glamorous, funny, or exciting images and catchy tunes. Write down a list of photos and tunes from adverts that remind you of things you could buy. How is conditioning involved in making us more likely to buy these products?

Key words

conditioning
conditioned reflex

Find out about

▶ how human beings learn new things

▶ explanations that scientists have for how your memory works

G Human learning

When humans and other mammals experience something new, they can develop new ways of responding. Experience changes human behaviour, and this is called learning.

Intelligence, memory, consciousness, and language are complex functions carried out by the outer layer of the brain, which is called the cerebral cortex. These functions are all involved in learning.

How does learning happen?

Neurons in your brain are connected together to form complicated **pathways**. How do these pathways develop? The first time a nerve impulse travels along a particular pathway, from one neuron to another, new connections are made between the neurons. New experiences set up new neuron pathways in your brain.

If the experience is repeated, or the stimulus is particularly strong, more nerve impulses follow the same nerve pathway. Each time this happens, the connections between these neurons are strengthened. Strengthened connections make it easier for nerve impulses to travel along a pathway. As a result, the response you produce becomes easier to make.

The brains of human babies develop new nerve pathways very quickly. Your brain can develop new pathways all your life. This means you can still learn as you get older, though more slowly

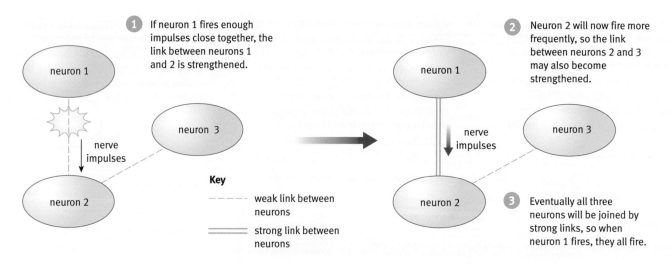

1 If neuron 1 fires enough impulses close together, the link between neurons 1 and 2 is strengthened.

2 Neuron 2 will now fire more frequently, so the link between neurons 2 and 3 may also become strengthened.

Key

- - - - weak link between neurons

===== strong link between neurons

3 Eventually all three neurons will be joined by strong links, so when neuron 1 fires, they all fire.

Nerve pathways form in a baby's brain as a result of a stimulus from its environment. Repeating the stimulus strengthens the pathway. The baby then responds in the same way each time it receives the stimulus. Some neurons in the brain do not take part in any pathway. Many of these unused neurons are destroyed.

Repetition

Repetition helps you learn because it strengthens the pathways the brain uses to carry out a particular skill. Perhaps you have learned to ride a bicycle, play a musical instrument, perform a new dance sequence or touch type. To do these things you created new nerve pathways then strengthened them through repetition.

This made it easy for you to respond in the way that you practised.

For example, Marie is a gymnast. When she has to learn new movements she stands still and imagines going through the motions – the position of her body and muscles being used at each stage. Visualization works because thinking about using a muscle triggers nerve impulses to that muscle. This strengthens the pathways the impulse takes. After a period of visualization, the actual movement is a lot easier to perform.

Marie visualizes new movements to help her learn them.

Age and learning

You learn to speak through repetition because you are surrounded by people talking. Children learn language extremely easily up to the age of about eight years. Their brains easily make new neuron pathways in the language processing region. As we get older it becomes harder for this part of the brain to make new pathways.

Feral children

In 1799, in southern France, a remarkable creature crept out of the forest. He acted like an animal but looked human. He could not talk. The food he liked and the scars on his body showed he had lived wild for most of his life. He was a wild, or **feral**, child. The local people guessed that he was about twelve years old and named him Victor.

Victor was taken to Paris. He lived with a doctor who tried to tame him and teach him language. At first people thought Victor had something wrong with his tongue or voice box. He could only hiss when people tried to teach him the names of objects. He communicated in howls and grunts.

Victor never learned to say more than a couple of words. By the time he was found, the time in his development when it was easy to learn language had passed.

Key words

neurons
pathways
repetition
feral

Questions

1 Write a few sentences to explain how you learn by experience Use the key word on this page in your answer

2 Explain why repeating a skill helps you learn it.

3 Write a list of skills you could practise by visualization.

H What is memory?

Psychologists are scientists who study the human mind. They describe **memory** as your ability to store and retrieve information.

Short-term and long-term memory

Read this sentence:

▶ As you read this sentence you are using your **short-term memory**.

Short-term memory lasts for about 30 seconds in most people. If you have no short-term memory you will not be able to make sense of this sentence. By the time you get to the end of the sentence, you will have forgotten the beginning.

Think about a song you know the words to:

▶ To remember the words you use your **long-term memory**.

Verbal memory is *any information* you store about words and language. It can be divided into short and long-term memory. Long-term memory is a lasting store of information. There seems to be no limit to how much can be stored in long-term memory. And the stored information can last a lifetime.

Different memory stores work separately

People with advanced Alzheimer's disease suffer short-term memory loss. They cannot remember what day it is, or follow simple instructions. But they may still remember their childhood clearly.

Some people lose long-term memory because of brain damage or disease. Their short-term memory is normal. This evidence is important because it shows that long-term and short-term memory must work separately in the brain.

The 'Nun Study' at the University of Kentucky, USA, has had the participation of 678 School Sisters of Notre Dame. They ranged in age from 75 to 106 years. The sisters have allowed scientists to assess their mental and physical function every year and to examine their brains at death. The study has led to significant advances and discoveries in the area of Alzheimer's disease and other brain disorders.

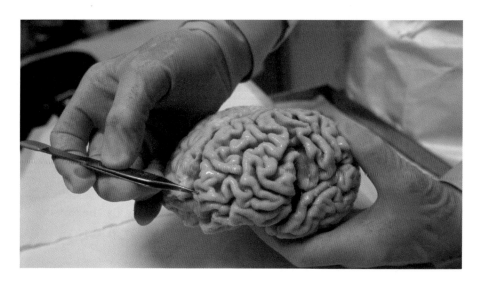

Sensory memory store

You can also use a sensory memory store to store sound and visual information for a short time. When you wave a sparkler on bonfire night it leaves a trail of light. You can even write shapes in the air that other people can see. You see the trail because you store each image of the sparkler separately for a short time. The ability to store images for a short time makes the separate pictures in a film seem continuous. You can store sound temporarily in the same way.

Your sensory memory stores each image of the sparkler separately for a short time. This makes the whole shape seem continuous.

Questions

1 Write down one sentence to describe memory.

2 What is the difference between short-term and long-term memory?

3 Explain why a person with advanced Alzheimer's disease is unable to do simple things like go shopping or cook for themselves.

4 Give one piece of evidence that short-term and long-term memory are separate.

Key words

memory
short-term memory
long-term memory

How much can you store in your short-term memory?

Cover up the list of letters below with a piece of paper. Move the paper down so you can see just the top row. Read through the row once, then cover it up and try to write down the letter sequence. Then go down to the next row and do the same. Find out how many letters you can remember in the correct order.

```
N T

A N L

N F E K

B F E X A

N A Z T P L

M B T F E Q P

U N D A C X Z G

O R B V E X Z D A

R T L D C A G P V E
```

If you remembered more than seven letters in a row correctly, then you have excellent short-term memory. Short-term memory can only store about seven items. When you are remembering letters in a list, each letter is an 'item'. To remember more letters, chunk them into groups.

For example, the row O R B V E X Z D A has nine letters. Chunk these into groups of three: 'ORB' 'VEX' 'ZDA'.

The nine letters are easy to remember because now they are only three items. Three items doesn't overload your short-term memory

Models of memory

Trying to remember word lists is a way of testing your memory. Memory tests can tell us what memory can and cannot do. But they do not explain how the neurons in the brain work to give you memory. Explanations for how memory happens are called **models of memory**.

Key words

models of memory
multi store model

The multi-store model: memory stores work together

Read through the list of words below once. Then cover the page and try to write down as many of the words as you can remember. They can be in any order.

> dog, window, film, menu, archer, slave, lamp, coat, bottle, paper, kettle, stage, fairy, hobby, package

How many did you remember? If this type of test is carried out on large numbers of people, a pattern is seen in the words they recall. People often remember the last few words on the list and get more of them right. They also recall the first few words on the list quite well.

When you look at a list of words:

▶ Nerve impulses travel from your eyes to your sensory memory.

▶ Some sensory information is passed on to your short-term memory. Only the information you pay attention to is passed on. You will not be able to remember words you have not noticed.

▶ If more information arrives than the short-term memory can hold then some is lost (forgotten). You will not remember these words either.

▶ Some information is passed to your long-term memory. These are words you will remember – usually the first few words on the list.

▶ The last information your short-term memory receives will still be there when you start to write down the list. So these are also words you will remember usually the last few words on the list.

This use of sensory, short-term and long-term memory is known as the **multi-store model of memory**.

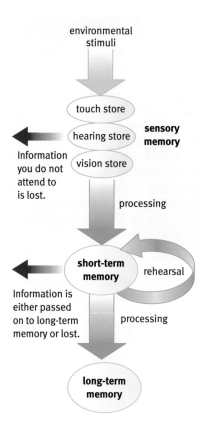

The multi-store model of memory can be used to explain how some information is passed to the long-term memory store and some information is lost.

> **Questions**
>
> **5** You read the menu on a board inside a café. When you try to tell your friend sitting outside all the choices, you forget some. Why can you not remember everything on the list?
>
> **6** 'So far none of the models of memory can explain completely how your memory works.' What must an explanation do to be accepted by scientists?

Rehearsal is one technique actors use to learn their lines

Rehearsal and long-term memory

Look at this row of letters:

R T I D A C G P E V

There are too many letters in this row for you to store them separately in your short-term memory. Given time you would probably repeat the letters over and over until you remembered them. Rehearsal is a well-known way of memorizing things. An actor can memorize a sonnet (a fourteen-line poem) in around 45 minutes. Psychologists think that rehearsal moves information from your short-term memory to your long-term memory store.

The working-memory model

In 1972, two psychologists, Fergus Craik and Robert Lockhart concluded that the multi-store memory model was too simple. They suggested that rehearsal is only one way to transfer information from short-term to long-term memory.

Rehearsed information is processed and stored rather than lost from short-term memory. Craik and Lockhart argued that you are more likely to remember information if you process it more deeply. They suggested that if you understand the information, or it means something to you, you will process it more deeply.

For example, if you can see a pattern in the information, you process it more deeply. So:

AAT, BAT, CAT, DAT, EAT

is much easier to remember than:

DAT, AAT, EAT, CAT, BAT

You also process information more deeply if there is a strong stimulus linked to the information, for example, colour, smell, or sound.

An active working memory

Short-term memory is now seen as an active '**working memory**'. Here you can hold and process information that you are consciously thinking about. Communication between long-term and working memory is in both directions. This way you can retrieve information you need, and also store information you may need later.

Putting it into practice

You can apply what the psychologists have discovered to your own school work.

- *Repetition:* If you are struggling to remember a piece of information you have read, read it several times.

- *Rehearsal:* Read sections of what you have to learn that are short enough to keep in your short-term memory. Make notes from memory to help move the information to your long-term memory.

- *Active memory:* Use highlighter pens to pick out key facts. Add to your old notes as you learn new topics.

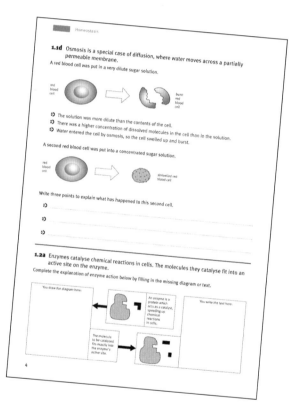

Ben is doing homework. He uses both his long-term and working memory.

> **Key words**
>
> working memory

Questions

7 Make two lists of 10 different things to buy from a supermarket. Try to remember one list. Put the second list into 'families' e.g. tins, bakery, cleaning, and memorize it. Which list is easier to remember and why?

8 Give an example of something you can remember because a strong stimulus is linked to the information:

   a a colour **b** a smell **c** a sound

9 Write down an example of something you have memorized through rehearsal, for example the directions to the cinema, or a complicated set of moves in a computer game. How has rehearsal helped you to remember this?

10 Explain why using a highlighter pen to pick out key facts makes your revision more successful.

Summary

In this module you have explored how your nervous system controls your responses to changes in the environment. These may be simple responses or involve more complex behaviour.

A mammal's nervous system

▷ The nervous system is made up of the central nervous system (CNS) and peripheral nervous system.

▷ The CNS coordinates an animal's responses via sensory and motor neurons.

▷ Receptors and effectors can be single cells or parts of complex organs.

Nerve impulses

▷ The structure of a neuron means that it can carry electrical impulses.

▷ Neurons are separated by tiny gaps called synapses.

▷ Some drugs and toxins affect the transmission of impulses across synapses.

▷ Impulses pass across synapses via chemical transmitters, which bind to specific receptor molecules. H

Reflex actions

▷ Living things respond to stimuli to help them survive.

▷ Simple reflexes travel through reflex arcs.

▷ Animals that only have simple behaviour are at a disadvantage because they cannot respond to new situations.

▷ Conditioned reflexes increase an animal's chance of survival. H

▷ The brain can modify some reflex responses. H

The brain

▷ The human brain has billions of neurons.

▷ New neuron pathways form during development.

▷ During learning some pathways become more likely to transmit impulses than others, so some skills may be learnt through repetition.

▷ The variety of potential pathways in the brain means that mammals can adapt to new situations.

▷ Some learning can only happen at particular times in development. H

Memory

▷ Scientists have used different methods to map the brain's cerebral cortex.

▷ Memory is the storage and retrieval of information.

▷ Memory can be divided into short-term and long-term memory.

▷ Scientists have suggested several models to explain memory, but so far none of them can account for all the observations.

▷ Humans are more likely to remember information if they can see a pattern in it. H

Questions

1 A car is waiting at the traffic lights. As soon as the light turns green, the driver's leg muscle moves his foot. He presses on the accelerator and the car moves forward.

   a Name the stimulus, the receptor, and the effector in this response.

   b As the driver accelerates, some dust blows into the car, making him sneeze. Explain why sneezing is an example of a reflex action.

2 All living things use reflexes. Simple animals use reflexes to control all their behaviour. But more complex animals also make conscious decisions.

   a Name two reflexes seen in human babies.

   b Name two reflexes also seen in human adults.

   c Explain why reflexes are important for survival. Use these key words in your answer: *stimulus, involuntary, response*

   d Explain the disadvantage of having only reflex behaviour.

3 A person stands on a drawing pin. Pain receptors in the skin detected the stimulus. They quickly move their foot upwards.

   a How does this reflex help the person to survive?

   b Draw a diagram to describe this reflex arc. Label the diagram with notes to explain what is happening at each stage. (The central nervous system in this reflex is the spinal cord.)

4 a Draw a labelled diagram of a motor neuron.

   b Describe two jobs done by the fatty sheath.

   c Name the gap between two neurons.

   d Explain how nerve impulses pass from one neuron to the next. [H]

5 Many different kinds of nasty-tasting caterpillars have black-and-yellow markings. [H]

   a Explain how birds learn to avoid eating prey with black-and-yellow markings. [H]

   b Suggest what might happen if a nice-tasting caterpillar had black-and-yellow markings. [H]

6 Your brain has billions of neurons. It allows you to learn from experience.

   a Explain how you learnt a new skill, e.g. riding a bike. Include these key words in your answer: *neuron pathway, impulses, repetition*

   b Suggest why it is easier to learn some skills at a certain age. [H]

7 a Name four functions of the cortex in the human brain.

   b What methods have scientists used to find out this information?

8 a What memory are you using to:

   i read this sentence?

   ii remember the words of a song?

   b Scientists have come up with several models to describe how memory works.

   i Describe one of these models.

   ii Explain why none of the models scientists have come up with so far is a full explanation of memory.

Why study chemical synthesis?

We use chemicals to preserve food, treat disease, and decorate our homes. Many of these chemicals are synthetic. Developing new products, such as drugs to treat disease, depends on the chemists who synthesize and test new chemicals.

The science

Chemists who synthesize new chemicals need knowledge of science explanations combined with practical skills. It is important to understand how to control reactions so that they are neither too slow nor dangerously fast. Planning a synthesis involves calculating how much of the reactants to mix together to make the amount of product required.

Acids are important reagents in synthesis. Ionic theory can explain the characteristic behaviours of these chemicals.

Chemistry in action

Synthesis provides many of the chemicals that we need for food processing, health care, cleaning and decorating, modern sporting materials, and many other products. The chemical industry today is developing new processes for manufacturing these chemicals more efficiently and with less impact on the environment.

Chemical synthesis

Find out about:

- the importance of the chemical industry
- the steps involved in the synthesis of a new chemical
- techniques for controlling the rate of chemical change
- a theory to explain acids and alkalis
- ways to measure the efficiency of chemical synthesis

Find out about:

▶ the chemical industry
▶ bulk and fine chemicals
▶ the importance of chemical synthesis

The chemical industry converts raw materials into pure chemicals which are then used in synthesis to make a wide range of products.

A The chemical industry

The chemical industry converts raw materials, such as crude oil, natural gas, minerals, air and water, into useful products. The products include chemicals for use as dyes, food additives, fertilizers, dyestuffs, paints, and pharmaceutical drugs.

The industry makes **bulk chemicals** on a scale of thousands or even millions of tonnes per year. Examples are ammonia, sulfuric acid, sodium hydroxide, chlorine, and ethene.

On a much smaller scale, the industry makes **fine chemicals** such as drugs and pesticides. It also makes small quantities of speciality chemicals needed by other manufacturers for particular purposes. These include such things as flame retardants, food additives, and the liquid crystals for flat-screen televisions and computer displays.

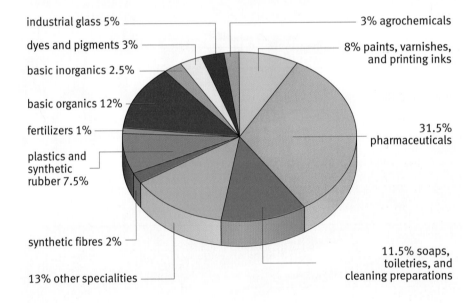

industrial glass 5%
dyes and pigments 3%
basic inorganics 2.5%
basic organics 12%
fertilizers 1%
plastics and synthetic rubber 7.5%
synthetic fibres 2%
13% other specialities

3% agrochemicals
8% paints, varnishes, and printing inks
31.5% pharmaceuticals
11.5% soaps, toiletries, and cleaning preparations

The pie chart shows the range of products made by the chemical industry in Britain and their relative value.

The part of a chemical works which produces a chemical is called a plant. Some of the chemical reactions occur at a high temperature, so that a source of energy is needed. Also, a lot of electric power is need for pumps to move reactants and products from one part of the plant to another. Sensors monitor the conditions at all the key points in the plant. The data is fed to computers in the control centre, where the technical team controls the plant.

Key words

bulk chemicals
fine chemicals
scale up

People in the chemical industry

People with many different skills are needed in the industry. Research chemists work in laboratories to find new processes and develop new products.

The industry needs new processes so that it can be more competitive and more sustainable. The aim is to use smaller amounts of raw materials and energy while creating less waste.

People devising new products have to work closely with people in the marketing and sales department. They are able to say if the novel product is wanted. If the new product is promising it may first be tried out by making it in a pilot plant.

As part of the market research, possible new products are given to customers for trial. At the same time, financial experts estimate the value of the new product in the market. They then compare this with the cost of making the product to check that the new process will be profitable.

Chemical engineers have to **scale up** the process and design a full-scale plant. This can cost hundreds of millions of pounds.

Some chemicals from the industry go directly on sale to the public, but most of them are used to make other products. Transport workers carry the chemicals to the industry's customers.

Every chemical plant needs managers and administrators to control the whole operation. There are also people in service departments look after the needs of the people working there which includes medical and catering staff, training, and safety officers.

Questions

1 Give the name and chemical formula of a bulk chemical.

2 What percentage value of products of the chemical industry in Britain are used:
   a in agriculture and horticulture?
   b to make polymers?
   c for medical diagnosis and treatment?

3 List these chemicals under two headings: 'bulk chemical' and 'fine chemical'.
   ‣ the drug aspirin
   ‣ the hydrocarbon ethene
   ‣ the perfume chemical citral
   ‣ the acid sulfuric acid
   ‣ the herbicide glyphosate
   ‣ the alkali sodium hydroxide
   ‣ the food dye carotene
   ‣ the pigment titanium dioxide

Plant operators monitor the processes from a control room.

Maintenance workers help to keep the plant running.

Find out about:
- acids and alkalis
- the pH scale
- reactions of acids

B Acids and alkalis

Acids

The word **acid** sounds dangerous. Nitric, sulfuric, and hydrochloric acids are very dangerous when they are concentrated. You must handle them with great care. These acids are less of a hazard when diluted with water. Dilute hydrochloric acid, for example, does not hurt the skin if you wash it away quickly, but it stings in a cut and rots clothing.

Not all acids are dangerous to life. Many acids are part of life itself. Biochemists have discovered the citric acid cycle. This is a series of reactions in all cells. The cycle harnesses the energy from respiration for movement and growth in living things.

Organic acids

Organic acids are molecular. They are made of groups of atoms. Their molecules consist of carbon hydrogen and oxygen atoms. The acidity of these acids arises from the hydrogen in the —COOH group of atoms.

Acetic acid (chemical name: ethanoic acid) is the acid in vinegar. Most white vinegar is just a dilute solution of acetic acid. Brown vinegars have other chemicals in the solution that give the vinegar its colour and flavour. Most micro-organisms cannot survive in acid, so vinegar is used as a preservative in pickles (E260). The pure acid is a liquid.

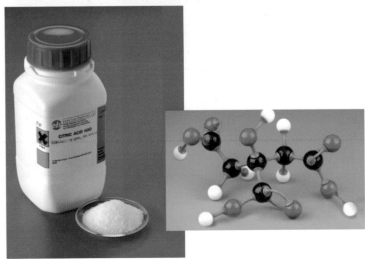

Citric acid is found in citrus fruits like oranges and lemons. The human body turns over about 2 kg of citric acid a day during respiration. The acid is manufactured on a large scale for the food industry. Citric acid (E330) and its salts (E331 and E332) are added to food to prevent them reacting with oxygen in the air – they are antioxidants. They also give a tart taste to drinks and sweets.

Mineral acids

Sulfuric, hydrochloric, and nitric acids come from inorganic or mineral sources. The pure acids are all molecular. Sulfuric and nitric acids are liquids at room temperature. Hydrogen chloride is a gas which becomes hydrochloric acid when it dissolves in water.

Alkalis

Pharmacists sell antacids in tablets to control heartburn and indigestion. The chemicals in these medicines are the chemical opposites of acids. They are designed to neutralize excess acid produced in the stomach – hence the name 'antacids'.

Some chemical antacids are soluble in water to give a solution with a pH above 7. Chemists call them **alkalis**. Common alkalis are sodium hydroxide, NaOH; potassium hydroxide, KOH; and calcium hydroxide, $Ca(OH)_2$.

The traditional name for sodium hydroxide is 'caustic soda'. The word caustic means that the chemical attacks living tissue including skin. Alkalis can do more damage to delicate tissues than dilute acids. Caustic alkalis are used in the strongest oven and drain cleaners. They clearly have to be used with great care.

Sulfuric acid, H_2SO_4, is manufactured from sulfur, oxygen, and water. The pure, concentrated acid is an oily liquid. The chemical industry in the UK makes about 2 million tonnes of the acid each year. The acid is essential for the manufacture of other chemicals, including detergents, pigments, dyes, plastics, and fertilizers.

The old name for hydrogen chloride was 'spirits of salt'. It forms when concentrated sulfuric acid is added to salt crystals. Hydrogen chloride, HCl, is a gas which fumes in moist air and is very soluble in water. Today, the chemical industry makes most of the hydrogen chloride it needs as a by-product of the production of other chemicals, such as the polymer PVC.

Oven cleaners often contain caustic alkalis.

Key words

acid
alkalis

Questions

1 Work out from the pictures of molecules the formulae of:
 a acetic acid
 b tartaric acid

2 What are the formulae of these antacids?
 a Magnesium hydroxide made up of magnesium ions, Mg^{2+}, and hydroxide ions, OH^-.
 b Aluminium hydroxide made up of aluminium ions, Al^{3+}, and hydroxide ions, OH^-.

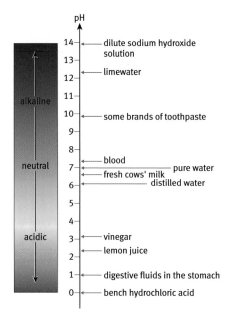

pH

- 14 — dilute sodium hydroxide solution
- 13
- 12 — limewater
- 11
- 10 — some brands of toothpaste
- 9
- 8
- 7 — blood — pure water
- fresh cows' milk
- 6 — distilled water
- 5
- 4
- 3 — vinegar
- 2 — lemon juice
- 1 — digestive fluids in the stomach
- 0 — bench hydrochloric acid

alkaline

neutral

acidic

The pH scale

Reactions of acids

The reactions of acids are important not just to chemists but to everyone living in a consumer society.

Acids with indicators

The term pH appears on many cosmetic, shampoo and food labels. It is a measure of acidity. The **pH scale** is a number scale which shows the acidity or alkalinity of a solution in water. Most laboratory solutions have a pH in the range 1–14.

Indicators change colour to show whether a solution is acidic. Litmus turns red in acid solution. Special mixed indicators, such as universal indicator, show a range of colours and can be used to estimate pH values.

Acids with metals

Acids react with **metals** to produce **salts**. The other product is hydrogen gas.

$$\text{acid} + \text{metal} \longrightarrow \text{salt} + \text{hydrogen}$$

Not all metals will react in this way. You may remember the list of metals in order of reactivity in Module C4 *Chemical patterns*, page 137. Metals below lead in the list do not react with acids, and even with lead it is hard to detect any change in a short time.

Etching metal

One method of chemical etching is based on the reaction of acids with metals. Etching is a way of producing multiple copies of printed pictures that have the quality of original drawings.

- First a metal plate of zinc or steel is covered with wax.

- Then the artist scrapes the wax with a stylus to make the drawing.

- The plate is then dipped in an acid bath. The acid reacts only with the metal that has been exposed. The metal still coated with wax is protected and so does not react.

- The artist uses a feather to brush away hydrogen bubbles that stick to the plate. If they were left on the plate the acid would not 'bite' into that spot.

- When the acid has removed enough metal, the plate is taken from the acid and rinsed. The remaining wax is removed.

- The plate is now ready to be covered with ink, which flows into the grooves etched by the acid.

Using a feather to brush away hydrogen bubbles while etching a metal plate with acid

Acids with metal oxides or hydroxides

An acid reacts with a **metal oxide** or **hydroxide** to form a salt with water. No gas forms.

$$\text{acid} + \text{metal oxide (or hydroxide)} \longrightarrow \text{salt} + \text{water}$$

The reaction between an acid and a metal oxide is often a vital step in making useful chemicals from ores.

Acids with carbonates

Acids react with **carbonates** to form a salt, water, and bubbles of carbon dioxide gas. Geologists can test for carbonates by dripping hydrochloric acid onto rocks. If they see any fizzing, the rocks contain a carbonate. This is likely to be calcium carbonate or magnesium carbonate.

$$2HCl(aq) + CaCO_3(aq) \longrightarrow CaCl_2(aq) + H_2O(l) + CO_2(g)$$

This is a foolproof test for the carbonate ion. So the term 'the acid test' has come to be used to describe any way of providing definite proof.

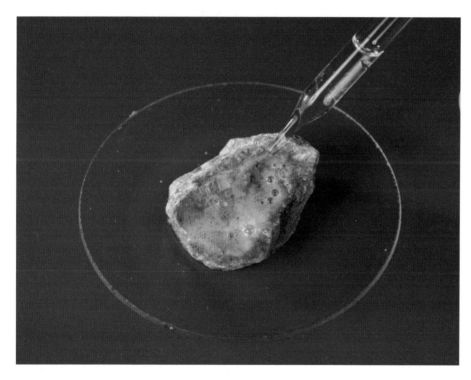

Testing for limestone using hydrochloric acid

Key words

pH scale
indicators
metals
salts
metal oxide
metal hydroxide
carbonates

Questions

3 Write a balanced equation for the reaction which takes place when a steel plate is etched with hydrochloric acid. The salt formed is iron(II) chloride.

4 Magnesium hydroxide, $Mg(OH)_2$, is an antacid used to neutralize excess stomach acid, HCl. Write a balanced equation for the reaction.

5 There is a volcano in Tanzania, Africa, whose lava contains sodium carbonate, Na_2CO_3. The cooled lava fizzes with hydrochloric acid. Write a balanced equation for the reaction.

6 Limescale forms in kettles, boilers, and pipes where hard water is heated. Limescale consists of calcium carbonate. Three acids are often used to remove limescale: citric acid, acetic acid (in vinegar), and dilute hydrochloric acid. Which acid would you use to de-scale an electric kettle and why?

c Salts from acids

What makes an acid an acid?

Chemists have a theory to explain why all the different compounds that are acids behave in a similar way when they react with indicators, metals, carbonates, and metal oxides.

It turns out that acids do not simply mix with water when they dissolve. They react, and when they react with water they produce hydrogen ions. For example, hydrochloric acid is a solution of hydrogen chloride in water. The HCl molecules react with the water to produce **hydrogen ions** and chloride ions.

$$HCl(g) + water \longrightarrow H^+(aq) + Cl^-(aq)$$

The theory of acids is an ionic theory. Any compound is an acid if it produces hydrogen ions when it dissolves in water.

All acids contain hydrogen in their formula. Nitric acid, HNO_3, and phosphoric acid, H_3PO_3 both contain hydrogen. But not all chemicals that contain hydrogen are acids. Ethane, C_2H_6, and ethanol, C_2H_5OH, are not acids.

In organic acids it is only the hydrogen atoms in the —COOH groups that can ionize when the acid dissolves in water.

Strong and weak acids

Some acids ionize completely when they dissolve in water. Chemists call them strong acids. Hydrochloric, sulfuric and nitric acids are strong acids.

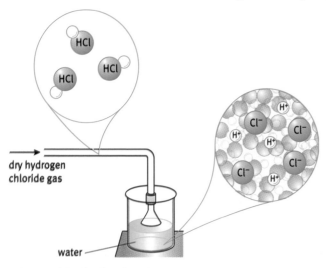

dry hydrogen chloride gas

water

Hydrogen chloride dissolves in water to make hydrochloric acid. The HCl molecules react with water to form ions

Organic acids only ionize slightly when they dissolve. They are weak acids. In vinegar, for example, only about one in a hundred of the ethanoic acid molecules are ionized. This helps to explain why vinegar is pH 3 but dilute hydrochloric acid is pH 1.

What makes a solution alkaline?

Alkalis such as the soluble metal hydroxides are ionic compounds. They consist of metal ions and **hydroxide ions**. When they dissolve, they add hydroxide ions to water. It is these ions which make the solution alkaline.

$$NaOH(aq) + water \longrightarrow Na^+(aq) + OH^-(aq)$$

Neutralization

Sodium hydroxide and hydrochloric acid react to produce a salt (sodium chloride) and water.

$$Na^+(aq) + OH^-(aq) + H^+(aq) + Cl^-(aq) \longrightarrow Na^+(aq) + Cl^-(aq) + H_2O(l)$$

During a **neutralization reaction** the hydrogen ions from an acid react with hydroxide ions from the alkali to make water.

$$H^+(aq) + OH^-(aq) \longrightarrow H_2O(l)$$

The remaining ions in the solution make a salt.

Salts

Salts form when a metal oxide, or hydroxide, neutralizes an acid. So every salt can be thought of as having two parents. Salts are related to a parent metal oxide or hydroxide and to a parent acid.

Salts are ionic (see Module C4 *Chemical patterns*, Section J). Most salts consist of a positive metal ion combined with a negative non-metal ion. The metal ion comes from the parent metal oxide or hydroxide. The non-metal ion comes from the parent acid.

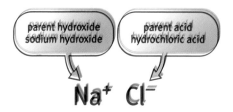

parent hydroxide
sodium hydroxide

parent acid
hydrochloric acid

Na⁺ Cl⁻

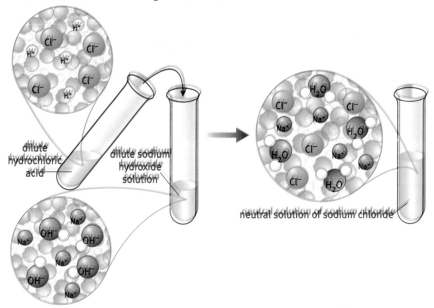

dilute hydrochloric acid

dilute sodium hydroxide solution

neutral solution of sodium chloride

Dilute sodium hydroxide solution neutralizes dilute hydrochloric acid, forming a neutral solution of sodium chloride.

It is possible to work out the formulae of salts knowing the charges on the ions (see Module C4 *Chemical patterns*, page 59). Some non-metal ions consist of more than one atom. The table below includes some examples. In the formula for magnesium nitrate, $Mg(NO_3)_2$, the brackets around the NO_3 show that two complete nitrate ions appear in the formula.

Non-metal ions that consist of more than one atom	Symbols
carbonate	CO_3^{2-}
hydroxide	OH^-
nitrate	NO_3^-
sulfate	SO_4^{2-}

Questions

1 Write equations to show what happens when these compounds dissolve in water:
   a nitric acid
   b sulfuric acid
   c calcium hydroxide

2 Identify the parent acid and a possible parent metal oxide or hydroxide that can react to form:
   a lithium chloride
   b calcium nitrate
   c magnesium sulfate

3 Use the tables of ions on page 59 and on this page to write down the formulae of these salts:
   a potassium nitrate
   b magnesium carbonate
   c sodium sulfate
   d calcium nitrate

Find out about:
- everyday uses of salts
- salts used in dialysis

D Salts in our lives

Soluble salts find their way into many areas of our lives. It is the job of chemists to make them for us. In Module C5 *Chemicals of the natural environment*, you learned about how most of the starting materials for the stuff we use in our daily lives comes from the lithosphere. These raw materials have to be transformed into the right chemicals.

The reactions of acids with metals, oxides, hydroxides, and carbonates can all be used to make valuable salts. For uses such as food or medicines, these salts have to be made pure.

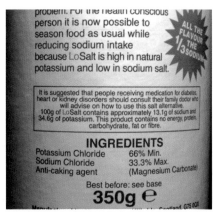

Potassium chloride, KCl, is used by people who are trying to cut down the amount of sodium in their diet, very often to help reduce high blood pressure. It is also added to fertilizers to provide the potassium ions that plants need for growth.

Potassium nitrate, KNO_3, is used in curing meat to make things like bacon and pastrami. It is also used as fertilizer and as an important constituent of gunpowder and fireworks.

Sodium benzoate, $C_7H_5O_2Na$, is widely used as a food preservative (E211). It is used to prevent bacteria spoiling food. The salt works best to preserve food if the pH is low, and so it is added to foods such jams, salad dressing, fruit juices, pickles, and carbonated drinks.

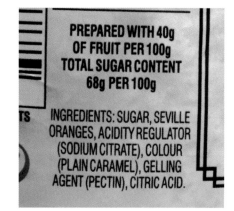

Monosodium citrate, $C_6H_7O_7Na$, is a food additive (E331) which helps to control the pH of foods and to make an antioxidant more effective. So it prevents food going off in the presence of air.

Making calcium chloride for dialysis

The kidneys remove toxic chemicals from blood. In cases of kidney failure, patients are put on dialysis machines that do the job outside the body. Blood passes out of the body through a tube into the dialysis machine.

Inside the machine the blood goes past a special membrane. On the other side of the membrane is a solution containing a mixture of salts at the same concentrations as the same salts in the blood. The toxic chemicals pass from the blood through the membrane into the solution. It is also possible for salts to pass back into the blood.

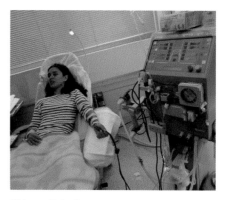

Kidney dialysis

One of the salts in the dialysis solution is calcium chloride. It is a particularly important salt as the level of calcium in the blood has to be maintained at a particular level. Just a little bit too much or too little and the patient could become very ill indeed. The calcium chloride therefore has to be very pure, and the quantity added to the solution has to be measured accurately.

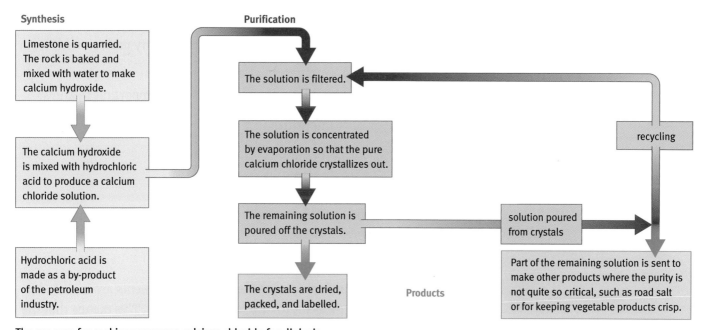

The process for making very pure calcium chloride for dialysis

---

Questions

1 Identify the parent metal oxide or hydroxide and a possible parent acid that can react to form:
   a potassium chloride
   b potassium nitrate
   c sodium citrate

2 Refer to the flow diagram for the process to make calcium chloride.
   a Write the balanced equation for the reaction used to make the salt.

   b Identify steps taken to make the yield of the pure salt as large as possible.

3 Chemical food additives are given E numbers. E331 can be monosodium citrate, disodium citrate, or trisodium citrate. With the help of the model shown on page 208 suggest an explanation for the fact that citric can form three sodium salts.

E Purity of chemicals

Grades of purity

Suppliers of chemicals offer a range of grades of chemicals. In a school laboratory you might use one of these grades: technical, general laboratory, and analytical. The purest grade is the analytical grade.

CALCIUM CARBONATE PRECIPITATED CP
CAS No: 471-34-1
EC No: 207-469-9
QTY: 1kg   BNO: C1042/R6 - 708717

Assay	99%
Chloride (Cl)	0.005%
Sulphate (SO₄)	0.05%
Iron (Fe)	0.002%
Lead (Pb)	0.002%

Label on a bottle of laboratory grade calcium carbonate

Calcium carbonate, for example, is used in a blast furnace to extract iron from its ores. It is also an ingredient of indigestion tablets. The iron industry can use limestone straight from a quarry. Limestone has some impurities but they do not stop it from doing its job in a blast furnace.

The calcium carbonate in an indigestion tablet must be safe to swallow. It must be very pure.

Purifying a chemical is done in stages. Each stage takes time and money, and becomes more difficult. So the higher the purity, the more expensive the chemical. Manufacturers therefore buy the quality most suitable for their purpose.

When deciding what grade of chemical to use, it is important to know:

▶ the amount of impurities

▶ what the impurities are

▶ how they can affect the process

▶ whether they will end up in the product, and whether it matters if they do

Testing purity

Medicines contain an active ingredient. Other ingredients are included to make the medicine pleasant to taste and easy to take. This means that the pharmaceutical companies that make medicines need sweeteners, food flavours, and other additives.

The companies buy in many of their ingredients. Technical chemists working for the companies have to make sure that the suppliers are delivering the right grade of chemical. The aim is to make sure that all the chemicals are between 99% and 100% pure.

Citric acid is widely used to control the pH in syrups such as cough medicines. Technicians can check the purity of the acid using a procedure called a **titration**. It is possible to calculate the purity of a sample of citric acid by measuring the volume of alkali that it can exactly neutralize. The analyst has to know the precise concentration of the alkali.

Steps involved in a titration

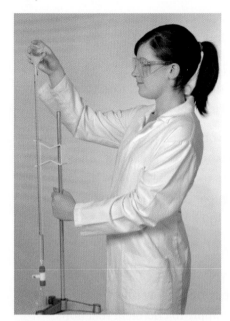

1 The technician fills a **burette** with a solution of sodium hydroxide. She knows the concentration of the alkali.

2 The technician weighs out accurately a sample of citric acid and dissolves it in water.

3 The technician dissolves the acid in pure water. Then she adds a few drops of phenolphthalein indicator. The indicator is colourless in the acid solution.

4 The technician adds alkali from the burette. She swirls the contents of the flask as the alkali runs in. Near the end she adds the alkali drop by drop. At the **end point** all the citric acid is just neutralized. The indicator is now permanently pink.

Questions

1 Make a list of some uses of sodium chloride (salt). Beside each use write which grade of salt it would be best to use.

2 Epsom salts consist of magnesium sulfate. Magnesium sulfate is soluble in water. Produce a flow diagram to show how you could remove impurities that are insoluble in water from a sample of Epsom salts. Processes you might use include: crystallization, dissolving, drying, evaporation, filtration.

Find out about:

▶ measuring rates of reaction
▶ factors affecting rates of reaction
▶ catalysts in industry
▶ collision theory

An explosion is an example of a very fast chemical reaction

F Rates of reaction

Controlling reaction rates

Some chemical reactions seem to happen in an instant. An explosion is an example of a very fast reaction.

Other reactions take time – seconds, minutes, hours or even years. Rusting is a slow reaction and so is the rotting of food.

Chemists synthesizing chemicals must work out a procedure that is as efficient as possible. It is important that the chosen reactions happen at a convenient speed. A reaction that occurs too quickly can be hazardous. A reaction that takes several days to complete is not practical because it ties up equipment and people's time for too long.

Measuring rates of reaction

Your pulse rate is the number of times your heart beats every minute. The production rate in a factory is a measure of how many articles are made in a particular time. Similar ideas apply to chemical reactions.

Chemists measure the **rate of a reaction** by finding the quantity of product produced or the quantity of reactant used up in a fixed time.

For the reaction

$$Mg(s) + 2HCl(aq) \longrightarrow MgCl_2(aq) + H_2(g)$$

the rate can be measured quite easily by collecting and measuring the hydrogen gas.

$$\text{average rate} = \frac{\text{change in the volume of hydrogen}}{\text{time for the change to happen}}$$

In most chemical reactions the rate changes with time. The graph on the left is a plot of the volume of hydrogen formed against time for the reaction of magnesium with acid. The graph is steepest at the start showing that the rate of reaction was greatest at that point. As the reaction continues the rate decreases until the reaction finally stops. The steepness of the line is a measure of the rate of reaction.

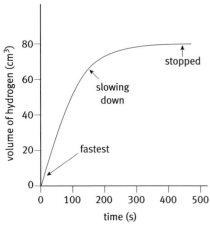

A plot of the volume of hydrogen formed against time for a reaction of magnesium with hydrochloric acid

Key words

rate of reaction

Questions

1 For each of these reactions, pick from page 219 a method that could be used to measure the rate of reaction:
   a $CaCO_3(aq) + 2HCl(aq) \longrightarrow CaCl_2(aq) + CO_2(g) + H_2O(l)$
   b $Zn(s) + H_2SO_4(aq) \longrightarrow ZnSO_4(aq) + H_2(g)$
   c $Na_2S_2O_3(aq) + 2HCl(aq) \longrightarrow 2NaCl(aq) + SO_2(aq) + S(s) + H_2O(l)$

Methods of measuring rates of reaction

Collecting and measuring a gas product

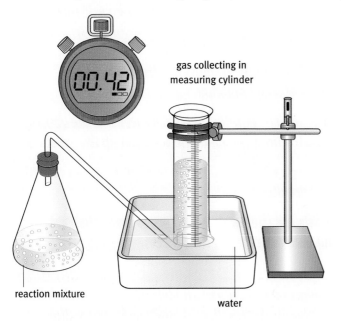

gas collecting in measuring cylinder

reaction mixture

water

Record the volume at regular intervals, such as every 30 or 60 seconds.

Measuring the loss of mass as a gas forms

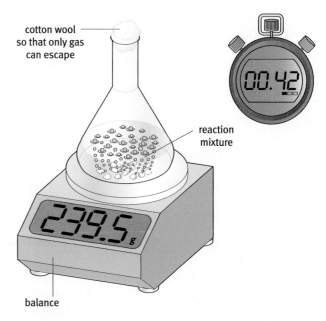

cotton wool so that only gas can escape

reaction mixture

balance

Record the mass at regular intervals such as every 30 or 60 seconds.

Timing how long it takes for a small amount of solid reactant to disappear

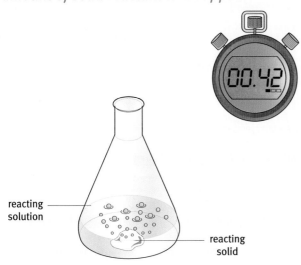

reacting solution

reacting solid

Mix the solid and liquid in the flask and start the timer. Stop it when you can no longer see any solid.

Timing how long it takes for a solution to turn cloudy

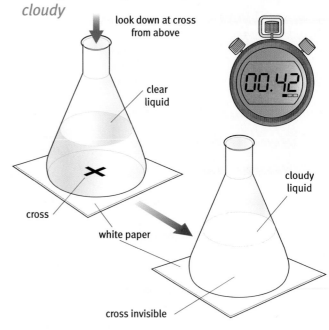

look down at cross from above

clear liquid

cross

white paper

cloudy liquid

cross invisible

This is for reactions that produce an insoluble solid. Mix the liquids in the flask and start the timer. Stop it when you can no longer see the cross on the paper through the solution.

Factors affecting reaction rates

A sliced loaf goes stale faster than an unsliced loaf. Milk standing in a warm kitchen goes sour more quickly than milk kept in a refrigerator. Changing the conditions alters the rate of these processes and may others.

Factors which affect the rate of chemical reactions are:

▶ The *concentration* of reactants in solution – the higher the concentration the faster the reaction.

▶ The *surface area* of solids – powdering a solid increases the surface area in contact with a liquid or solid and so speed up the reaction.

▶ The *temperature* – typically a 10 °C rise in temperature can roughly double the rate of reaction.

▶ *Catalysts* – catalysts are chemicals which speed up a chemical reaction without being used up in the process.

The factors in action

The apparatus in the diagram was used in an investigation into the effect of changing the conditions on the reaction of zinc metal with sulfuric acid. The graph below shows the results.

$$Zn(s) + H_2SO_4(aq) \longrightarrow ZnSO_4(aq) + H_2(g)$$

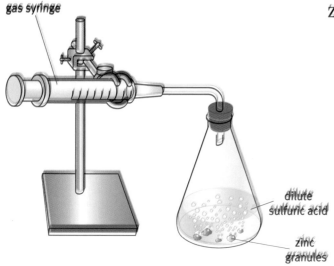

Apparatus used to investigate the factors affecting the rate of reaction of zinc with sulfuric acid.

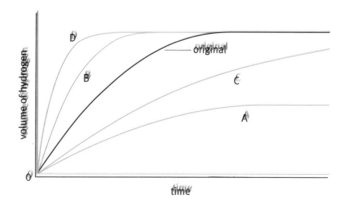

Plots showing the volume of hydrogen formed over time during an investigation of the factors affecting the rate of reaction of zinc with sulfuric acid. The investigator used the same mass of zinc each time. There was more than enough metal to react with all the acid.

The red line on the graph plots the volume of hydrogen gas against time using zinc granules and 50 cm^3 of dilute sulfuric acid at 20 °C. The reaction gradually slows down and stops because the acid concentration falls to zero. There is more than enough metal to react with all the acid. The zinc is in excess.

The effect of concentration

Line A on the graph shows the results of repeating the procedure using acid that was half as concentrated while leaving all the other conditions the same.

The investigator added 50 cm³ of this more dilute acid. Halving the acid **concentration** lowers the rate at the start. Again the reaction slows down because the concentration falls as the acid reacts with the metal. The final volume of gas is cut by half because there was only half as much acid to start with.

The effect of temperature

Line B shows the result of carrying out the reaction at 30 °C while leaving all the other conditions the same as in the original set up. This speeds up the reaction and more or less doubles the rate at the start. The quantities of chemicals are the same so the final volume of gas collected at room temperature is the same as it was originally.

The effect of surface area

Line C shows the result of keeping to all the original conditions but using the same excess of zinc metal in larger pieces. Fewer larger lumps of metal have a smaller **surface area** so the reaction starts more slowly. The amount of acid is unchanged and the metal is still in excess so that the final volume of hydrogen is the same.

one big lump several small lumps

Breaking up a solid into smaller pieces increases the surface area. This increases the amount of contact between the solid and the solution, making it possible for the reaction to go faster.

The effect of adding a catalyst

Line D shows what happens when the investigation is repeated with everything the same as in the original set up but with a **catalyst** added. Adding a few drops of copper(II) sulfate solution produces this effect. The reaction starts more quickly and the graph is steeper. Catalysts do not change the final amount of product, so the volume of gas at the end is the same as before.

Questions

2 How would you account for the fact that:
  a sliced bread goes stale more quickly than unsliced bread?
  b there is a danger of explosions in flour mills?

3 How is it possible to control conditions to slow down or stop these changes:
  a the rusting of iron?
  b a chip-pan fire?
  c milk going sour?

4 How is it possible to control conditions to speed up these changes:
  a the setting of an epoxy glue?
  b the cooking of an egg?
  c the conversion of oxides of nitrogen in car exhausts to nitrogen?
  d the conversion of sugar to alcohol and carbon dioxide?

Catalysts in industry

What is a catalyst?

A catalyst is a chemical that speeds up a chemical reaction. It takes part in the reaction, but is not used up.

Modern catalysts can be highly selective. This is important when reactants can undergo more than one chemical reactions to give a mixture of products. With a suitable catalyst it can be possible to speed up the reaction that gives the required product, but not speed up other possible reactions that create unwanted by-products.

Better catalysts

Catalysts are essential in many industrial processes. They make many processes possible economically. This means that chemical products can be made at a reasonable cost and sold at affordable prices.

Research into new catalysts is an important area of scientific work. This is shown by the industrial manufacture of ethanoic acid (see page 208) from methanol and carbon monoxide. This process was first developed by the company BASF in 1960 using a cobalt compound as the catalyst at 300 °C and at a pressure 700 times atmospheric pressure.

About six years later the company Monsanto developed a process using the same reaction, but a new catalyst system based on rhodium compounds. This ran under much milder conditions: 200 °C and 30–60 times atmospheric pressure.

In 1986, the petrochemical company BP bought the technology for making ethanoic acid from Monsanto. They have since devised a new catalyst based on compounds of iridium. This process is faster and more efficient. Iridium is cheaper, and less of the catalyst is needed. Iridium is even more selective so the yield of methanol is greater and there are fewer by-products. This makes it easier to make pure methanol and there is less waste.

The manufacture of ethanoic acid from methanol and carbon monoxide is only possible in the presence of a catalyst.

methanol + carbon monoxide → ethanoic acid

$$CH_3OH(g) + CO(g) \rightarrow CH_3COOH(g)$$

Reaction conditions:
pressure: 30 atmospheres
temperature: 200 °C
catalyst: iridium.

Collision theory

Chemists have a theory to explain how the various factors affect reaction rates. The basic idea is that molecules can only react if they bump into each other. Imagining molecules colliding with each other leads to a theory which can account for the effects of concentration, temperature and catalysts on reaction rates.

According to **collision theory**, when molecules collide some bonds between atoms can break while new bonds form. This creates new molecules.

Molecules are in constant motion in gases and liquid and there are millions upon millions of collisions every second. Most reactions would be explosive if every collision led to reaction. It turns out that only a very small fraction of all the collisions are 'successful' and actually lead to reaction. These are the collisions in which the molecules are moving with enough energy to break bonds between atoms.

Any change that increases the number of 'successful' collisions per second has the effect of increasing the rate of reaction.

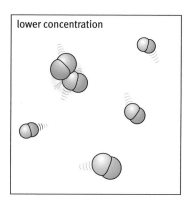
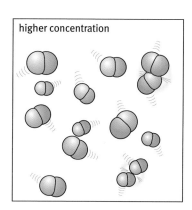

Molecules have a greater chance of colliding in a more concentrated solution. More collisions means more reaction. Reactions get faster if the reactants are more concentrated.

Questions

5 a Where do cobalt, rhodium and iridium appear in the periodic table (see Module C4, page 39).

   b Why is not surprising that these three metals can be used to make catalysts for the same process?

6 Suggest reasons why it is important to develop industrial processes which:

   a run at lower temperatures and pressures

   b produce less waste

Key words

collision theory

Find out about:
▶ reacting masses
▶ yields from chemical reactions

Key words

relative formula mass
reacting mass
actual yield
theoretical yield
percentage yield

G Chemical quantities

Chemists wanting to make a certain quantity of product need to work out how much of the starting materials to order. Getting the sums right matters, especially in industry, where a higher yield for a lower price can mean better profits.

The trick is to turn the symbols in the balanced chemical equation into masses in grams or tonnes. This is possible given the relative masses of the atoms in the periodic table (see Module C4 *Chemical patterns*, page 39).

Reacting masses

Adding up the relative atomic masses for all the atoms in the formula of a compound gives the **relative formula mass** of chemicals (see Module C5 *Chemicals of the natural environment*, page 139). Given the relative formula masses it is then possible to work out the masses of reactants and products in a balanced equation. These are the **reacting masses**.

Questions

1 What mass of:
   a copper(II) oxide reacts with 98 g H_2SO_4 in dilute sulfuric acid?
   b HCl in hydrochloric acid reacts with 100 g calcium carbonate?
   c HNO_3 in dilute nitric acid neutralizes 56 g of potassium hydroxide?

RULES FOR WORKING OUT REACTING MASSES

STEP 1   Write down the balanced symbol equation.

STEP 2   Work out the relative formula mass of each reactant and product.

STEP 3   Write the relative reacting masses under the balanced equation, taking into account the numbers used to balance the equation.

STEP 4   Convert to reacting masses by adding the units (g, kg, or tonnes).

STEP 5   Scale the quantities to amounts actually used in the synthesis or experiment.

Example

What are the masses of reactants and products when sulfuric acid reacts with sodium hydroxide?

Step 1 $2NaOH + H_2SO_4 \longrightarrow Na_2SO_4 + 2H_2O$

Step 2   relative formula mass of NaOH = 23 + 16 + 1 = 40

relative formula mass of H_2SO_4 = (2 × 1) + 32 + (4 × 16) = 98

relative formula mass of Na_2SO_4 = (2 × 23) + 32 + (4 × 16) = 142

relative formula mass of H_2O = (2 × 1) + 16 = 18

Steps 3 & 4

2NaOH	+	H_2SO_4	→	Na_2SO_4	+	$2H_2O$
2 × 40 = 80		98		142		2 × 18 = 36
80 g		98 g		142 g		36 g

Yields

The yield of any synthesis is the quantity of product obtained from known amounts of starting materials. The **actual yield** is the mass of product after it is separated from the mixture and purified and dried.

Theoretical yield

The **theoretical yield** is the mass of product expected if the reaction goes exactly as shown in the balanced equation. This is what could be obtained in theory if there are no by-products and no losses while chemicals are transferred from one container to another. The actual yield is always less than the theoretical yield.

Example

What is the theoretical yield of ethanoic acid made from 8 tonnes methanol? (See page 222.)

Step 1   *Write down the balanced equation*

$$\text{methanol} + \text{carbon monoxide} \longrightarrow \text{ethanoic acid}$$
$$\underset{32}{CH_3OH(g)} + \quad CO(g) \quad \longrightarrow \underset{60}{CH_3COOH(g)}$$

Step 2   *Work out the relative formula masses*

methanol: $12 + 4 + 16 = 32$

ethanoic acid: $24 + 4 + 32 = 60$

Steps 3 & 4   *Write down the relative reacting masses and convert to reacting masses by adding the units*

Theoretically, 32 tonnes of methanol should give 60 tonnes of ethanoic acid.

Step 5   *Scale to the quantities actually used*

If the theoretical yield of ethanoic acid = x tonnes, then

$$\frac{\text{mass of ethanoic acid}}{\text{mass of methanol}} = \frac{60 \text{ tonnes}}{32 \text{ tonnes}} = \frac{x \text{ tonnes}}{8 \text{ tonnes}}$$

So, the yield of ethanoic acid from 8 tonnes of methanol should be

$$8 \text{ tonnes} \times \frac{60 \text{ tonnes}}{32 \text{ tonnes}} = 15 \text{ tonnes}$$

Percentage yield

The **percentage yield** is the percentage of the theoretical yield that is actually obtained. It is always less than 100%.

Questions

2 What is the mass of salt which forms in solution when:

a hydrochloric acid neutralizes 4 g sodium hydroxide,

b 12.5 g zinc carbonate reacts with excess sulfuric acid?

3 A preparation of sodium sulfate began with 8.0 g of sodium hydroxide.

a Calculate the theoretical yield of sodium sulfate from 8.0 g sodium hydroxide.

b Calculate the percentage yield given that the actual yield was 12.0 g.

Example

What is the percentage yield if 8 tonnes of methanol produces 14.7 tonnes of ethanoic acid?

From the previous example:

$$\underset{\text{yield}}{\text{theoretical}} = 15 \text{ tonnes}$$

actual yield = 14.7 tonnes

$$\underset{\text{yield}}{\text{percentage}} = \frac{\text{actual yield}}{\text{theoretical yield}} \times 100$$

$$= \frac{14.7 \text{ tonnes}}{15 \text{ tonnes}} \times 100$$

$$= 98\%$$

Find out about:

▶ steps in synthesis
▶ making a soluble salt

An operator emptying magnesium sulfate into the tank of a sprayer on a farm. As well as being a micronutrient needed for healthy plant growth, the salt is needed as a:
▶ raw material in soaps and detergents
▶ laxative in medicine
▶ refreshing additive in bath water
▶ raw material in the manufacture of other magnesium compounds
▶ supplement in feed for poultry and cattle
▶ coagulant in the manufacture of some plastics

H Stages in chemical synthesis

Chemical synthesis is a way of making new compounds. Synthesis puts things together to make something new. It is the opposite of analysis which takes things apart to see what they are made of.

The process of making a soluble salt on a laboratory scale illustrates the stages in a chemical synthesis.

In the following method an excess of solid is added to make sure that all the acid is used up. This method is only suitable if the solid added to the acid is either insoluble in water or does not react with water.

Making a sample of magnesium sulfate

Choosing the reaction

Any of the characteristic reactions of acids can all be used to make salts:

▶ acid + metal ⟶ salt + hydrogen

▶ acid + metal oxide or hydroxide ⟶ salt + water

▶ acid + metal carbonate ⟶ salt + carbon dioxide + water

Magnesium metal is relatively expensive because it has to be extracted from one of its compounds. So it makes sense to use either magnesium oxide or carbonate as the starting point for making magnesium sulfate from sulfuric acid.

Carrying out a risk assessment

It is always important to minimize exposures to risk. You should take care to identify hazardous chemicals. You should also look for hazards arising from equipment or procedures. This is a **risk assessment**.

In this preparation the magnesium compounds are not hazardous. The dilute sulfuric acid is an irritant, which means that you should keep it off your skin and especially protect your eyes. You should always wear eye protection when handling chemicals, for example.

Working out the quantities to use

In this procedure the solid is added in excess. This means that the amount of product is determined by the volume and concentration of the sulfuric acid. The concentration of dilute sulfuric acid is 98 g/litre. It turns out that a volume of 50 cm³ dilute sulfuric acid is suitable. This contains 4.9 g of the acid.

Carrying out the reaction in suitable apparatus under the right conditions

The reaction is fast enough at room temperature, especially if the magnesium carbonate is supplied as a fine power.

This reaction can be safely carried out in a beaker. Stirring with a glass rod makes sure that the magnesium carbonate and acid mix well. Stirring also helps to prevent the mixture frothing up and out of the beaker.

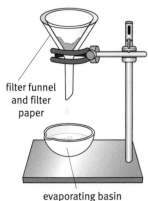

solid

stirring rod

dilute acid

1 Measure the required volume of acid into a beaker. Add the metal or the insoluble oxide, hydroxide or carbonate bit by bit until no more dissolves in the acid. Warm when most of the acid has been used up. Make sure that there is a slight excess of solid before moving on to the next stage.

Separating the product from the reaction mixture

Filtering is a quick and easy way of separating the solution of the product from the excess solid. The mixture filters faster if the mixture is warm.

filter funnel and filter paper

evaporating basin

2 Filter off the excess solid collecting the solution of the salt in an evaporating basin. The residue on the filter paper is the excess solid.

Questions

1 Write the balanced equation for the reaction of magnesium carbonate with sulfuric acid.

2 Why does the mixture of magnesium carbonate and sulfuric acid froth up?

3 What is the advantage of:
   a using powdered magnesium carbonate?
   b warming when most of the acid has been used up?
   c adding a slight excess of the solid to the acid?

4 Why is it impossible to use this method to make a pure metal sulfate by the reaction of dilute sulfuric acid with:
   a lithium metal?
   b sodium hydroxide?
   c potassium carbonate?

5 Look at the procedure on pages 227–228 described for making magnesium sulfate and identify risks that might arise from:
   a chemicals that react vigorously and spill over
   b chemicals that might spit or splash on heating
   c hot apparatus that might cause burns
   d apparatus that might crack and form sharp edges

Key words

risk assessment

Purifying the product

After the mixture has been filtered, the filtrate contains the pure salt dissolved in water. Evaporating much of the water speeds up crystallization. This is conveniently carried out in an evaporating basin. The concentrated solution can then be left to cool and crystallize.

Once the crystals are nearly dry they can be transferred to a desiccator. This is a closed container which contains a solid that absorbs water strongly.

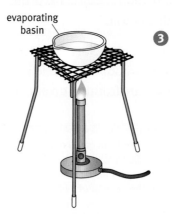

evaporating basin

3 Heat gently to evaporate some of the water. Evaporate until crystals form when a droplet of solution picked up on a glass rod crystallizes on cooling.

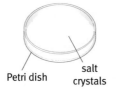

Petri dish   salt crystals

4 Pour the concentrated solution into a labelled Petri dish and set it aside to cool slowly.

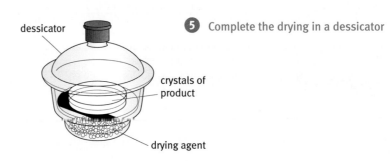

dessicator

crystals of product

drying agent

5 Complete the drying in a dessicator

Questions

6 Calculate the theoretical yield of magnesium sulfate crystals produced in the synthesis. (Include the water in the crystals when working out the relative formula mass of the salt.)

7 Identify the impurities removed during the purification stages.

8 Why is it important that the magnesium carbonate is added to the sulfuric acid in excess?

9 a What is the percentage yield of magnesium sulfate?

   b Suggest reasons why the percentage yield is less than 100% in this preparation.

Measuring the yield and checking the purity of the product

The final step is to transfer the dry crystals to a weighed sample tube and reweigh it to find the actual yield of crystals. Often it is important to carry out tests to check that the product is pure.

The appearance of the crystals can give a clue to the quality of the product. A microscope can help if the crystals are small. The crystals of a pure product are often well-formed and even in shape.

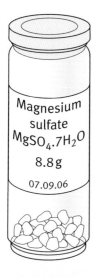

Magnesium
sulfate
$MgSO_4.7H_2O$

8.8 g

07.09.06

6 The weighed sample of product showing the name and formula of the chemical, the mass of product, and the date it was made.

Crystals of pure magnesium sulfate seen through a Polaroid filter (×60)

Summary

You have learnt how chemists can use their knowledge of chemical reactions to plan and carry out the synthesis of new compounds.

The chemical industry

▶ The chemical industry is an important part of the country's economy.

▶ Industry makes bulk chemicals on a large scale and fine chemicals for more specialized purposed on a smaller scale.

▶ Chemical synthesis provides useful products such as food additives, fertilizers, dyestuffs, paints, pigments, and pharmaceuticals.

The ionic theory of acids and alkalis

▶ Acids react in characteristic ways with metals, metal oxides, and metal carbonate.

▶ Alkalis neutralize acids to form salts.

▶ All acids have similar properties because they produce hydrogen ions, $H^+(aq)$, in water.

▶ Alkalis produce aqueous hydroxide ions, $OH^-(aq)$, when they dissolve in water.

▶ During a neutralization reaction, the hydrogen ions from an acid react with hydroxide ions from an alkali to make water.

Controlling the rate of change

▶ Chemists follow the rate of a change by measuring the disappearance of a reactant or the formation of a product.

▶ Factors which affect the rate of change include the concentration of reactants, the particle size of solid reactants, the temperature, and the presence of catalysts.

▶ Collision theory can be used to explain why changing the concentration of reactants affects the rate of reaction. H

Synthesis

▶ A chemical synthesis involves a number of stages, including:

   − choosing the reaction or series of reactions to make the required product

   − carrying out a risk assessment

   − working out the quantities of reactants to use H

   − carrying out the reaction in suitable apparatus in the right conditions

   − separating the product from the reaction mixture

   − purifying the product

   − measuring the yield and checking the purity of the product

▶ The balanced equation for the reaction is used to work out the quantities of chemicals to use and to calculate the theoretical yield.

▶ A titration is a technique that can be used to check the purity of chemicals used in synthesis.

Questions

1   Give examples to show why each of these products of chemical synthesis are useful and valuable:

   a food additives

   b fertilizers,

   c dyestuffs

   d paints

   e pharmaceuticals

2   Use the ionic theory of acids and alkalis to explain why:

   a Solutions of acids in water conduct electricity with hydrogen forming at the negative electrode.

   b Solutions of the hydroxides of lithium, sodium, and potassium are alkaline.

   c Water is one of the products of a neutralization reaction.

3   A small piece of metal is cut from a stick of lithium taken from the bottle of oil in which it is stored. the mass of the metal sample is 0.1 g. The piece is put into an apparatus containing excess water. After the metal and water have been allowed to mix, the volume of hydrogen produced is measured at intervals.

Time (minutes)	Volume of gas (cm³)
1	8
2	24
3	72
4	138
5	172
6	172

   a Why is lithium stored in oil?

   b Draw a diagram of an apparatus that could be used to collect and measure the hydrogen produced by the reaction. Show how it could be arranged so that the lithium and water do not mix until the experimenter is ready to start the reaction.

   c Plot the results on graph paper with labelled axes. Plot time on the horizontal axis.

   d What is the average rate of reaction between 3 and 4 minutes. The rate can be measured by the rate of formation of gas in cm³ per minute.

   e Explain the change in rate between 4 and 5 minutes.

   f It is suspected that that the results in the first 4 minutes are affected by some oil on the surface of the metal. Draw a dotted line on the graph to show the results that might be expected if the measurements are repeated using 0.1 g of lithium from which all traces of oil are removed.

4   Zinc sulfate is a soluble salt used in some dietary supplements. It can be made by reacting zinc oxide with dilute sulfuric acid. Zinc oxide is an insoluble white solid.

   These are the first steps in the synthesis of zinc sulfate:

   – Add small portions of powdered zinc oxide to some warm sulfuric acid. Stir the mixture after each addition. Keep doing this until the mixture is slightly cloudy.

   – Filter the mixture once all the acid has been used up.

   a Why is zinc oxide added until the mixture stays slightly cloudy?

   b Suggest reasons for using:

     i powdered zinc oxide

    ii warm sulfuric acid

   c What is the purpose of filtering the mixture?

   d Write a word equation for the reaction and then a balanced symbol equation.

   e Describe the next steps needed to obtain dry crystals of zinc sulfate from the liquid that passes through the filter paper.

5   Pure calcium chloride is used in kidney dialysis. One way of making the salt is to add an excess of powdered limestone (calcium carbonate) to dilute hydrochloric acid.

   a Calculate the theoretical yield of calcium chloride that can be made from 10 kg of calcium carbonate.

   b What is the percentage yield if the actual yield of calcium chloride is 9.9 kg from 10 kg calcium carbonate?

Why study waves?

A wave transfers energy through a medium, without the medium itself having to move as a whole. Scientists are interested in waves because many types of radiation behave in ways that are similar to waves we can see. Light and sound are two examples. Thinking of these as waves helps us explain and predict how they behave in different situations.

The science

When you disturb a 'springy' medium, the movement of one bit of the material makes the neighbouring bit move, after a slight delay. This causes a pulse to travel along. A wave is a continuous series of pulses moving through a medium. Waves are reflected when they hit a barrier, refracted when they cross from one medium into another, and are diffracted at corners and edges of obstacles. Two waves in the same region interfere. Light also has these properties. Light is one small part of the electromagnetic spectrum, a 'family' of waves whose properties depend on their frequency (or wavelength).

Physics in action

Physicists' understanding of waves has enabled them to develop new instruments, such as radio telescopes and infrared telescopes, for exploring the galaxy and beyond. Waves are also the foundation of modern communications, using radio and microwaves for television and radio broadcasts and for telephones, and optical fibres instead of wires to carry signals.

The wave model of radiation

Find out about:

- how vibrations can travel through a medium, and how this causes waves
- the speed at which a wave travels, its frequency, and its wavelength
- reflection, refraction, diffraction, and interference
- why scientists think of light and sound as waves
- the different radiations of the electromagnetic spectrum, how they differ, and how this affects their properties
- how waves of various kinds are used to carry information

Find out about:

▶ how waves travel through a medium

▶ the two main types of wave (transverse and longitudinal)

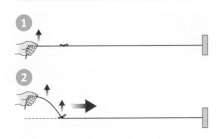

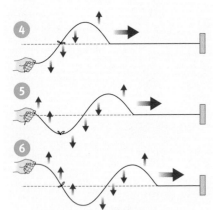

'Snap shots' of a pulse travelling along an elastic rope. The small arrows show how each bit of rope is moving. The large arrows show how the pulse moves.

Two meanings of 'medium'

Sometimes the word 'medium' is used to mean in the middle, between two extremes – like an average value. When we are talking about waves, the medium means the material the wave is travelling through. The plural of 'medium' is 'media'.

A   What is a wave?

A wave is a disturbance moving through a material. Imagine a long elastic rope lying on a flat table, with one end tied to a fixed point. If you move the other end from side to side, a pulse travels along the rope. The diagram on the left shows the shape of the rope at different instants.

What causes a wave?

You may have seen what happens if you topple a line of dominoes. As each one falls, it knocks the next one over – and so on, right down the line.

Toppling a line of dominos. The disturbance moves down the line, with a short delay from one domino to the next.

This is not a wave, however, just a single pulse. But now imagine that the dominoes are hinged at the bottom end, and connected to each other at the top by springs.

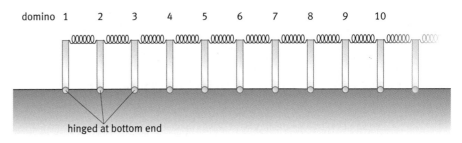

A line of dominoes hinged at the bottom and connected by springs at the top.

Think about what would happen if you moved the top of domino 1 to the right. As it moves, it compresses the first spring, which then pushes domino 2 to the right. This then compresses the second spring, pushing domino 3 to the right – and so on, along the line. The movement of domino 1 is repeated along the row, with a small time delay as it passes from one to the next. Now imagine moving the top of domino 1 slowly backwards and forwards. This sends a stream of pulses along the line – a continuous wave.

The source of a wave is always something that **vibrates**. The material that the wave travels through is called the **medium**. The medium has to be 'springy' and come back to its original position after being disturbed. In the example above, the line of dominoes and springs is the medium. For water waves, the medium is water. Gravity pulls the water back into position if it is disturbed.

Transverse and longitudinal

Two kinds of wave can travel through a medium. In one, the particles of the medium vibrate at right angles to the direction in which the wave moves. This is called a **transverse wave**. Water waves and waves on a rope are examples. But these can also be a wave where the particles of the medium vibrate in the same direction as the wave moves. Imagine a Slinky spring stretched across a table and fixed at one end. You can make a pulse by pushing the free end inwards, along the line of the spring, and back again. This compression pulse then travels along the spring. If you keep moving the free end in and out, this produces a continuous wave of pulses. This is called a **longitudinal wave**.

Key words
vibrates
medium
transverse wave
longitudinal wave

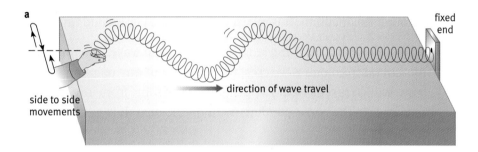

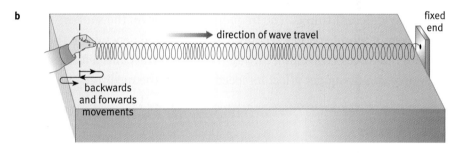

Two types of wave on a Slinky spring: **a** transverse; **b** longitudinal.

Sound is a longitudinal wave. A sound source vibrates, causing little compression pulses in the air nearby. Compressed air is 'springy'. As it recoils, it pushes on the neighbouring bit of air, which compresses it – and so on. A continuous wave of pulses travels through the air. Sound waves can travel through any gas, and also through liquids (you can hear underwater) and solids.

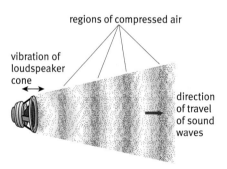

The vibrations of the loudspeaker cone send a stream of compression pulses through the air. This is a sound wave.

Questions

1 Look at the knot on the rope in the diagrams in the left margin.

   a Describe how it moves as the wave passes along the rope.

   b Sketch a graph to show how its displacement (its distance from the centre line) varies with time as the wave passes.

Find out about:

▶ how to describe a wave clearly
▶ how wave speed, frequency, and wavelength are connected

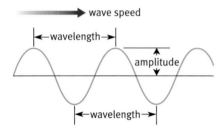

Important information when describing a wave

B Describing waves

To describe a wave clearly, there are several things you need to know:

1 Its **amplitude**. For water waves or waves on a rope or spring, this is the maximum distance that each point in the medium moves from its normal position as the wave passes. It is measured in metres (m).

2 Its **frequency**. This is the number of waves that pass any point in the medium every second. So it is the same as the number of vibrations per second of the source. Frequency is measured in hertz (Hz). 1 Hz means 1 wave per second.

3 Its **wave speed**. This is the speed at which each wave crest moves through the medium. It is measured in metres per second (m/s).

4 Its **wavelength**. This is the length of a complete wave. You might measure it from one wave crest to the next, but any point on the wave will do just as well. The distance is the same from any point on one wave to the corresponding point on the next. It is measured in metres (m).

It is important to realize that frequency and wave speed are two completely different things. The frequency depends on the source – how many times it vibrates every second. The wave speed depends on the medium the wave is travelling through. Once the wave has left the source, the source can no longer affect its speed through the medium.

Questions

1 Look at the diagrams of transverse and longitudinal waves on a Slinky on page 235. What would you change if you wanted to increase:

a the frequency of the transverse wave?

b the wave speed of the transverse wave?

c the amplitude of the longitudinal wave?

2 Estimate the wavelength of the sound waves in the diagram on page 235. Assume that the diagram is drawn to scale (life size).

Sound waves

For sound waves, the bigger the amplitude, the louder the sound. The greater the frequency, the higher the pitch of the sound. So, for example, the musical note C one octave above middle C has twice the frequency of middle C. The wave speed of sound (often just called the speed of sound) depends on the medium only. The bar graph below shows the speed of sound in some common media.

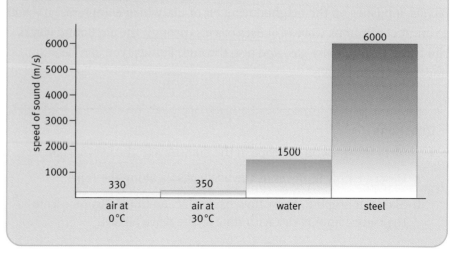

The wave equation

The frequency of a wave, its wave speed, and its wavelength are not independent of each other. There is a link between them, which applies to all waves of every kind. Imagine a source that vibrates five times per second. So it produces waves with a frequency of 5 Hz. If these have a wavelength of 2 metres in the medium they are travelling through, then every wave moves forward by 10 metres (5 × 2 m) in one second. The wave speed is 10 metres per second. In general,

$$\text{wave speed} \quad = \quad \text{frequency} \quad \times \quad \text{wavelength}$$

(metre per second, m/s)	(hertz, Hz)	(metre, m)
or (millimetre per second, mm/s)	(hertz, Hz)	(millimetre, mm)

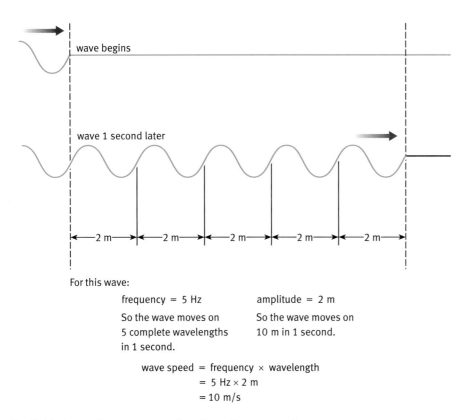

For this wave:

frequency = 5 Hz   amplitude = 2 m

So the wave moves on   So the wave moves on
5 complete wavelengths   10 m in 1 second.
in 1 second.

wave speed = frequency × wavelength
= 5 Hz × 2 m
= 10 m/s

The link between frequency, wavelength, and wave speed.

Is wave speed the same for all frequencies?

The frequency of a wave and its wave speed are two quite separate things. In fact the speed of sound waves in air is almost exactly the same for all audible frequencies. And the same is true of the speed of light in air. However, light waves of different frequencies travel at slightly different speeds in media like glass or water. This has some important consequences (as you will see later on page 243).

Questions

3 If you increase the frequency of waves through a medium, what will happen to their wavelength? Explain your reasoning.

4 When a stone is dropped into a pond, waves travel outwards at a speed of 500 mm/s. The wavelength of these waves is 100 mm. What is their frequency?

5 The speed of sound in air is 330 m/s. The note middle C has a frequency of 256 Hz. What is the wavelength of the sound wave when a musician plays this note? (You might find a calculator useful for this as the numbers do not work out neatly.)

6 If the speed of sound waves were very different for different frequencies, what effect would this have on the sounds we hear?

Find out about:
- four properties shared by all waves
- how waves behave at barriers, boundaries, edges, or obstacles

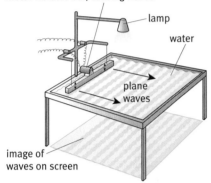

motor vibrates bar, causing waves

lamp

water

plane waves

image of waves on screen

A ripple tank producing a steady stream of plane waves.

Plane waves travelling at constant speed

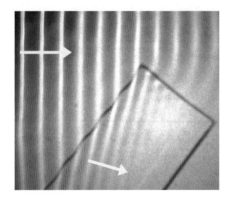

Refraction of water waves at a boundary between deep and shallow regions.

c Wave properties

Studying waves

Waves on a spring or a rope can travel in only one direction. Waves can be studied more fully by using a ripple tank. Waves are created on the surface of water and a lamp projects an image of the waves on to a screen.

Waves in water of constant depth are equally spaced. This shows that waves do not slow down as they travel. The wave speed stays the same. As the wave travels, it loses energy (because of friction). Its amplitude gets less but not its speed.

Using a ripple tank, you can observe four key properties of all waves.

Reflection

Water waves are reflected by a straight barrier placed in their path. If you draw a line at right angles to the barrier, the reflected waves make the same angle with this line as the incoming waves, but on the other side.

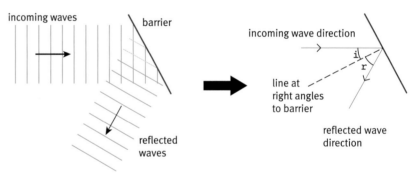

incoming waves

barrier

incoming wave direction

line at right angles to barrier

reflected waves

reflected wave direction

Reflection of water waves at a plane barrier. The angle of reflection (r) is equal to the angle of incidence (i).

Refraction

If waves cross a boundary from a deeper to a shallower region, they are closer together in the shallow region. The wavelength is smaller. This effect is called **refraction**. It happens because water waves travel slower in shallower water. The frequency (f) is the same in both regions, so the slower wave speed (v) means that the wavelength (λ) must be less (as $v = f\lambda$).

If the waves are travelling at an angle to the boundary between the two regions, their direction also changes. You can work out which way they will bend by thinking about which side of the wave gets slowed down first.

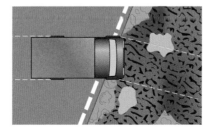

A model to help explain refraction. The truck goes more slowly on the muddy field. One wheel crosses the boundary first – so one side slows down before the other. This makes the truck change direction.

Diffraction

When water waves hit a barrier, they bend a little at the edge, and travel into the 'shadow' region behind the barrier. This effect is called **diffraction**. The longer the wavelength of a wave, the more it diffracts. At a gap between two barriers, waves bend a little at both edges. If the width of the gap is similar to their wavelength, the waves beyond the gap are almost perfect semicircles. If the gap is really tiny, much less than the wavelength of the waves, the waves do not go through at all.

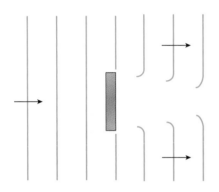

Diffraction occurs when waves meet an obstacle. The edges of the waves bend round the obstacle, into the shadow region behind.

Plane waves arrive at this harbour mouth. The waves inside the harbour are semicircular, because of diffraction at the narrow gap between the two piers.

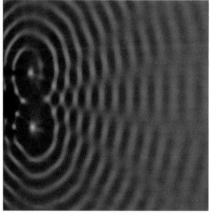

Waves diffracting at the edge of a barrier. Notice how some waves get round the corner.

Questions

1 Draw diagrams like the one on the left to show what happens as plane waves pass through a gap in a barrier that is:

   a bigger than their wavelength

   b about the same as their wavelength

Interference

When two waves meet, their effects add. If both waves have the same frequency, this causes an **interference** pattern. Where a crest of one wave meets a crest of the other, they add to make a bigger wavecrest. Where a crest of one meets a dip (or trough) of the other wave, the two cancel each other out.

The photograph on the right shows the surface of a ripple tank when two circular waves meet. Along some lines there is a large disturbance of the water surface. These are the points where the waves are 'in step' when they meet. Along other lines in between these, there is almost no disturbance of the water surface. Here the waves are 'out of step' when they meet, so they cancel each other out. The result is calm water.

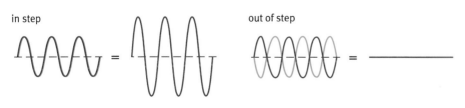

Two waves that are in step add to make a bigger disturbance. Two waves that are exactly out of step cancel each other out.

The interference pattern caused by circular waves spreading out from two dippers. Notice the lines of disturbed and calm water.

Key words

reflection

refraction

diffraction

interference

239

Find out about:

▶ a useful model for thinking about light and other radiations
▶ the evidence that light behaves like a wave

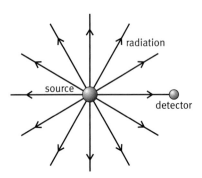

Radiation spreads out from a source and can affect another object (the detector) some distance away.

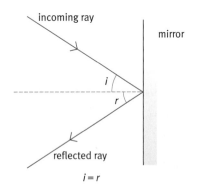

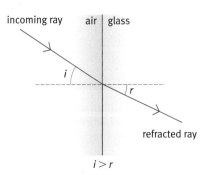

Light is reflected and refracted. This is consistent with light being a wave but does not really provide strong evidence, because a stream of particles would also do the same.

D Radiation and waves

Light is a type of **radiation**. Like all types of radiation, it travels out in all directions from a source until it hits another object. Here it might be **reflected** (bounce off), **transmitted** (go through), or **absorbed** (transfer energy to it), or a combination of all three. When radiation is absorbed, it causes changes in the absorber, which allow us to detect it.

There are many sources of light – electric light bulbs, candles, the Sun. Detectors include the retina of the eye, photographic film, light-dependent resistors (LDRs), and charge-coupled devices (CCDs) in digital cameras.

Light waves

People have always wondered what light really is. For a long time, scientists were unsure how best to think about light.

Isaac Newton (around 1664)

Thomas Young (in 1801)

Two of the wave properties – reflection and refraction – do not really provide conclusive evidence either way. A wave or a stream of particles would also be reflected by a barrier. And both would be refracted at a boundary where their speed changed.

The evidence that convinced scientists it was useful to think of light as a wave came from an experiment carried out by Thomas Young in 1801. Young used a narrow slit to produce a fine beam of light from a bright lamp.

He shone this on a slide with a double slit (two parallel clear lines on a black slide). On a screen about 1 metre away Young saw a pattern of bright and dark vertical lines. To understand this, look back at the photograph on page 239 of interference in a ripple tank. With water waves, there are lines of disturbance and lines of calm water. If the same happens with light, you would expect lines of brightness and lines of dimness, spreading out behind the double slit. On the screen, you should see a series of bright and dark patches.

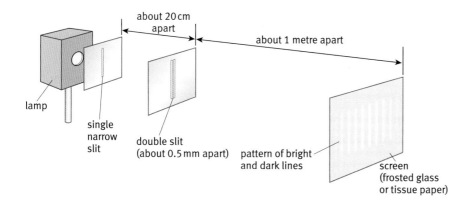

Young's experiment. The double slits are two sources of light of the same frequency. Light from these interferes to produce the pattern on the screen.

This is strong evidence that light behaves as a wave. If light were a stream of tiny particles, it could not produce an interference pattern in this way. Two streams of particles would not cancel each other if they met!

Young's experiment also involves diffraction. This the other property that is characteristic of waves. Light is diffracted by the two narrow slits. It spreads out into the region behind the slits, so that waves from the two sources overlap and interfere.

A wave model of radiation

If you look at water waves, you can see the medium (water) moving up and down. With sound waves, you cannot see the compression pulses in the medium, but you can detect them fairly easily. A candle flame in front of a loudspeaker flickers, showing the movements in the air around it. But with light, it is not possible to see or detect anything 'waving' or vibrating.

It is useful to think of light (and other radiations discussed later in this chapter) as a wave because it behaves in the same way as waves we can observe directly. These provide a useful model – a way of imagining what light is like. The wave model helps us explain our observations and predict what will happen in new situations.

What is the medium?

All the waves discussed earlier in this chapter need a medium to move through. But light can travel through a vacuum. It seems impossible to have a wave in a region where there is nothing at all. But in fact space (a vacuum) is not completely empty. Although it contains no matter, there can be fields in a vacuum. (Think, for example, of the gravitational field between the Sun and the Earth, which must act through empty space.) Scientists think that light is an **electromagnetic wave**. It consists of constantly varying electric and magnetic fields, which are able to move together through space.

Key words
radiation
reflected
transmitted
absorbed
electromagnetic wave

Questions
1 If you carefully covered one of the slits (of the double slit) in Young's experiment, what would you expect to see on the screen? Explain your answer.

2 What provides good evidence that sound is a wave?

Find out about:

▶ what happens to light crossing a boundary
▶ how optical fibres work
▶ the link between colour and frequency

E Bending light beams

Refraction and change of speed

A beam of light is refracted when it passes from one medium into another. This happens because the speed of light is different in different media. If the beam hits the boundary at an angle, one side is slowed down (or speeded up) before the other. This makes the beam change direction.

Total internal reflection

When a light beam crosses into a medium where it moves faster, it is bent towards the boundary between the media. By changing the direction of the ray in the first medium, you can make the emerging beam get closer to the boundary. At some point, it emerges right along the boundary. If you make the light hit the boundary at an even shallower angle, the boundary behaves as though it were a mirror. The light is reflected back into the first medium. This is called **total internal reflection (TIR)**.

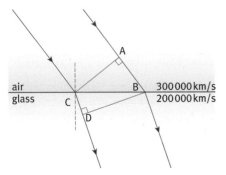

Light is refracted because its speed is different in the two media. Light travels the distance AB in air in the same time as it travels the distance CD in glass.

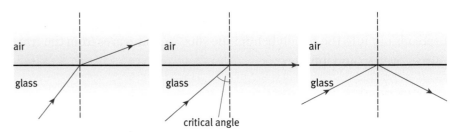

Light striking a boundary from glass to air at different angles.

Optical fibres

Optical fibres are at the forefront of the communications revolution. Thousands of telephone conversations are carried along a glass or clear plastic fibre, no thicker than a hair! An optical fibre makes use of total internal reflection. Light entering one end of a fibre is reflected repeatedly from the sides until it comes out the other end. Once inside the fibre, the light cannot escape as it always hits the surface at an angle that is bigger than the critical angle. It even follows bends in the fibre. The reflection is 'total', meaning that very little energy is lost through the sides. So the light can travel a long way down a fibre without getting much weaker.

Optical fibres enable doctors to look inside the body without using surgery. A bundle of optical fibres is used. Light is shone down some of the fibres to illuminate the object they want to inspect. A camera then takes a picture, using the light reflected from the object, coming back up the other fibres. Light from the object stays within the fibre it enters. So a clear image is visible at the other end.

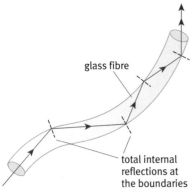

Light travels along an optical fibre because of total internal reflection.

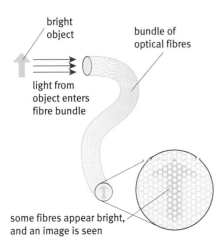

Using an optical fibre bundle to look at an inaccessible object.

Dispersion

A change of direction is not the only thing that happens to a light beam when it crosses a boundary between two transparent media. If a beam of white light passes through a triangular block (a prism), the emerging ray is coloured. This is called a **spectrum**. Newton carried out a famous series of experiments with prisms. He concluded that white light is really a mixture of the colours of the spectrum. This splitting of white light into colours is called dispersion.

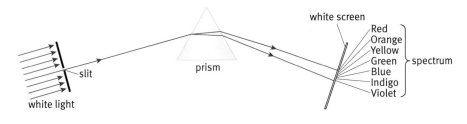

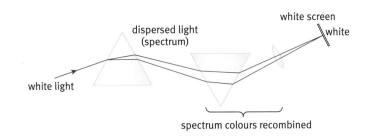

A second prism can recombine the coloured light rays to make white light again. So the colours are not caused by the glass of the prism.

Newton did not know what caused the different colours. Following Young's evidence that light is a wave (page 241), scientists were able to deduce that the colour of light depends on its frequency. (It therefore also depends on its wavelength, as the two are linked.) The visible band (waves that the human eye is able to detect) extends from red through to violet. Red has the lowest frequency (longest wavelength) and violet has the highest frequency (shortest wavelength).

Dispersion happens because light of different frequencies travels through glass (and other transparent media) at different speeds. The differences are small but they are enough to split the light up. In glass, red light travels faster than violet light. So it is not bent as much when it enters or leaves.

An optical fibre lamp. Light from a bulb inside the base travels through each fibre, so that each shows a tiny pinpoint of light at the end.

Key words

total internal reflection (TIR)
optical fibres
spectrum

Questions

1 These diagrams show rays of light travelling from one medium into another. Put the materials in order, from the one in which light travels slowest to the one in which it travels fastest.

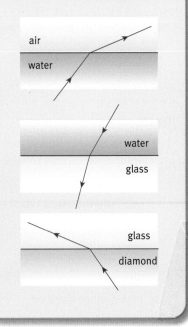

Find out about:

▶ the family of waves known as the electromagnetic spectrum

▶ the properties shared by all electromagnetic waves

F Electromagnetic waves

The human eye can detect light with frequencies between those of red and violet light. But there are other waves with frequencies bigger and smaller than this, which our eyes cannot detect. Visible light is just one member of a much larger family of electromagnetic waves – most of which cannot be detected by the human eye but require special equipment. These have several features in common:

▶ They can travel through empty space (a vacuum).

▶ They travel through space at a speed of 300 000 kilometres per second. This is usually called the **speed of light**, although it is the speed of all electromagnetic waves.

▶ They are transverse waves. However, it is tiny electric and magnetic fields that vibrate, rather than any material.

▶ They transfer energy. A source loses energy when it emits (sends out) electromagnetic waves. A material gains energy when it absorbs them.

Electromagnetic waves are very different from sound waves. Sound waves are longitudinal, travel much more slowly, and need a material (solid, liquid, or gas) to travel through.

The Sun emits electromagnetic waves, including infrared, light, and ultraviolet. All take the same time to reach the Earth because all travel through space at the same speed (see Question 4).

Photons

Around 1905, Albert Einstein and some other physicists carried out investigations which led them to think that electromagnetic radiation is always emitted and absorbed in 'packets', or **photons**.

▶ The higher the frequency of an electromagnetic wave, the more energy each photon has.

▶ The total amount of energy which a beam of radiation transfers each second to the absorber (its **intensity**) depends on:

  ▶ the energy of each photon in the beam

  ▶ the number of photons arriving every second

Key words

speed of light
photons
intensity
electromagnetic spectrum

Questions

1 *radio waves   sound waves   light*
  gamma rays   ultraviolet

  a Which of the above types of radiation is the odd one out, and why?

  b State three ways in which the other radiations are similar.

2 Name a type of electromagnetic radiation which

  a is visible to the eye

  b is emitted by hot objects

  c is used for radar

  d is emitted by radioactive materials

3 For each type of electromagnetic wave in the diagram on page 245, name one source and one detector.

4 The Sun is 150 000 000 km from Earth. Estimate how long it takes for the Sun's light to reach us.

The electromagnetic spectrum

Below, you can see the full range of electromagnetic waves. It is called the **electromagnetic spectrum**. At the bottom are the lowest-frequency radio waves. These have wavelengths of several kilometres. At the top are the highest-frequency gamma rays. These have wavelengths of less than a billionth of a millimetre.

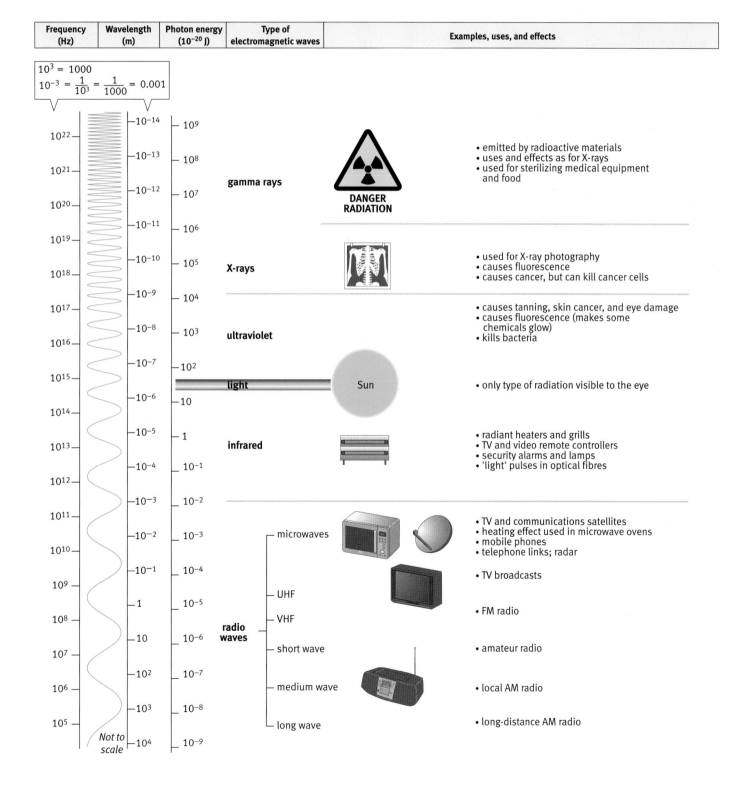

Frequency (Hz)	Wavelength (m)	Photon energy (10^{-20} J)	Type of electromagnetic waves	Examples, uses, and effects

$10^3 = 1000$
$10^{-3} = \dfrac{1}{10^3} = \dfrac{1}{1000} = 0.001$

gamma rays
- emitted by radioactive materials
- uses and effects as for X-rays
- used for sterilizing medical equipment and food

DANGER RADIATION

X-rays
- used for X-ray photography
- causes fluorescence
- causes cancer, but can kill cancer cells

ultraviolet
- causes tanning, skin cancer, and eye damage
- causes fluorescence (makes some chemicals glow)
- kills bacteria

light — Sun
- only type of radiation visible to the eye

infrared
- radiant heaters and grills
- TV and video remote controllers
- security alarms and lamps
- 'light' pulses in optical fibres

microwaves
- TV and communications satellites
- heating effect used in microwave ovens
- mobile phones
- telephone links; radar

- TV broadcasts

UHF
- FM radio

VHF

radio waves

short wave
- amateur radio

medium wave
- local AM radio

long wave
- long-distance AM radio

Not to scale

245

Find out about:

▸ electromagnetic waves with higher frequency than visible light
▸ ionizing radiations

G Above the visible

The human eye is sensitive to only a narrow range of frequencies within the electromagnetic spectrum. This is the range we call light. Together, the eye and brain sense the different frequencies as different colours. Violet light has the highest frequency and red the lowest. There are other kinds of electromagnetic radiation beyond the visible range.

Ultraviolet

Very hot objects, such as the Sun, emit some of their radiation beyond the violet end of the visible spectrum. This is **ultraviolet** radiation.

Most of the Sun's ultraviolet radiation does not reach the Earth's surface. High in the atmosphere, a band of gases called the ozone layer absorbs most of it. Only the 'near' ultraviolet, with frequencies closer to visible violet light, gets through. This is fortunate, as ultraviolet is harmful to living cells. If too much is absorbed by the skin, it can cause skin cancer. If you have black or dark skin, the ultraviolet is absorbed near the surface. But with fair skin, the ultraviolet can go deeper. Skin develops a tan to try to protect itself against ultraviolet. Ultraviolet can also damage the retina in the eye. On the other hand, ultraviolet is essential for the chemical reaction by which the body produces vitamin D. Vitamin D is needed by the immune system to resist diseases, including cancers.

Some materials fluoresce when they absorb ultraviolet. That is, they emit visible light and glow. Fluorescent paints look bright because they absorb the ultraviolet in sunlight and emit visible light. In fluorescent lamps the ultraviolet is produced by passing an electric current through the gas (mercury vapour) in the tube. The inside of the tube is coated with a white powder that emits light when it absorbs ultraviolet.

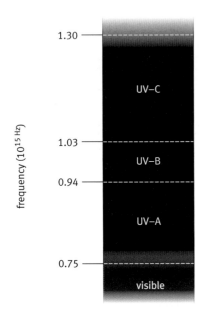

frequency (10^{15} Hz)

1.30 — UV–C

1.03 — UV–B

0.94 — UV–A

0.75 — visible

Ultraviolet is classified into three ranges according to its effect on skin. UV-C is the most dangerous but much of it is absorbed by ozone in the Earth's upper atmosphere. UV-A is not harmful in normal doses.

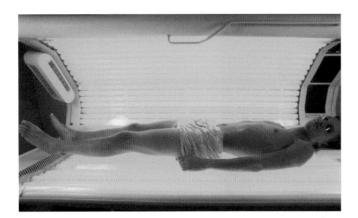

Sunbeds tan you because they emit UV-A.

Mars has a thin atmosphere and almost no ozone layer. Ultraviolet levels on its surface are much higher than on Earth. So simple life, such as bacteria, would find it very difficult to exist there.

Ionizing radiations

Ultraviolet and electromagnetic waves with still higher frequencies are ionizing. Each photon has enough energy to strip an electron from an atom in its path (remember: electrons are tiny charged particles in atoms). **Ionizing radiation** can alter the materials it strikes and, in the case of living cells, may damage or destroy them.

X-rays

X-rays have higher frequencies than ultraviolet. In an X-ray tube, X-rays are emitted when a beam of fast-moving electrons hits a metal target.

High-frequency X-rays are extremely penetrating. They can even pass through dense metals like lead. Lower-frequency X-rays are less penetrating. For example, they can pass through flesh but are absorbed by bone, so bones will show up on an X-ray photograph. X-rays are dangerous because they damage living cells deep in the body and can cause cancer. For this reason, exposure times are kept very short. However, concentrated beams of X-rays can be used to treat cancer by destroying abnormal cells.

At airports, X-ray machines are used for security checks on baggage. The wavelength is chosen so that the X-rays are absorbed by metal objects but pass through the softer, less-dense materials. So objects such as knives and guns can be seen on a screen.

Gamma rays

Gamma rays come from some radioactive materials. Most have shorter wavelengths than X-rays. However, the two types overlap (in frequency, and wavelength) in the electromagnetic spectrum, and there is no difference in behaviour between X-rays and gamma rays of the same wavelength. Like X-rays, gamma rays can be used in the treatment of cancer, and for taking X-ray-type photographs. As they can kill cells, they are used to kill harmful bacteria, for example to sterilize food and medical equipment.

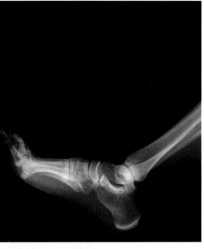

X-ray photograph. The X-rays pass through the flesh but not the bone. This is why these materials are seen as different shades in the picture.

Key words
ultraviolet
ionizing radiation
X-rays
gamma rays

Questions

1 The Sun emits many ultraviolet frequencies. Here are three of them:
0.8×10^{15} Hz 1.0×10^{15} Hz 1.2×10^{15} Hz

Which of these:
a is the most dangerous?

b is normally absorbed in the Earth's upper atmosphere?

c is safe in normal doses?

2 Someone claims to have invented a machine that produces X-rays so penetrating that they can pass right through any known material. Why would these X-rays be of no use for medical photos or security checks at airports?

3 **a** State two ways in which gamma rays are similar to X-rays.

b A key difference is in how two kinds of waves are produced. Explain what this difference is.

Find out about:

▶ electromagnetic waves with lower frequency than visible light

▶ how the wave model helps explain their behaviour

H Below the visible

Infrared

Beyond the red end of the spectrum there is radiation that the human eye cannot detect. This is **infrared** radiation.

When a radiant heater or grill is switched on, you can detect the infrared radiation coming from it. You notice the heating effect in your skin when it absorbs the radiation. In fact, all objects emit some infrared because of the motion of their atoms or molecules. Most emit a wide range of frequencies.

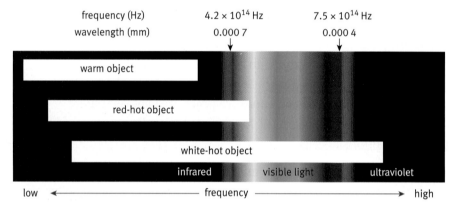

All objects emit some infrared. As the temperature rises, the wavelengths get shorter and enter the visible part of the spectrum.

As something heats up, it radiates more and more infrared, and at higher frequencies. At about 700 °C, the highest frequencies can be detectedby the eye, so the object glows 'red hot'. Above about 1000 °C, the emitted radiation includes the whole of the visible spectrum, so the object is 'white hot'.

Most solids absorb infrared: this makes their temperature rise. However, high-frequency infrared can pass through glass and clear plastics, although longer wavelengths are absorbed or reflected. The inside of a greenhouse heats up in the sunshine because the glass lets the Sun's light and high-frequency infrared through but reflects back the lower frequencies coming from the warmed materials inside. In other words, the energy is 'trapped'.

A greenhouse 'traps' energy by letting high-frequency infrared in but not letting low-frequency infrared out. This is sometimes called 'the greenhouse effect'.

Using infrared

Grills, toasters, and radiant heaters all use the heating effect of infrared. At night, warm things go on emitting infrared even though no light comes from them. Night-vision goggles and cameras detect this radiation and use it to produce a visible image. You can see an example on the left.

Most security lamps are switched on by motion sensors that detect the changing pattern of infrared caused by the warm body of an approaching person. Pulses of infrared can also transmit information. For example, a remote control uses them to send coded instructions to a TV. Infrared pulses can also be carried by optical fibres.

Although it is dark, the warm bodies of the people standing by the car and walking towards it are emitting infrared. This is detected by the camera and turned into a visual image.

Microwaves

Like all electromagnetic waves, **microwaves** have a heating effect when they are absorbed. This principle is used in microwave ovens, where microwaves penetrate deep into food and heat up the water in it.

In a microwave oven, the part that emits the radiation is called a magnetron. It is designed to emit microwaves with a frequency of 2450 MHz. This is lower than the frequency at which water absorbs best, but there is a good reason for this. If the absorption were much better, the microwaves would just heat the surface of the food, and not penetrate deep inside.

Microwaves are not ionizing radiation, however. A microwave photon does not have enough energy to knock an electron out of an atom.

Microwave communications

Microwaves have the highest frequencies (and shortest wavelengths) of all radio waves. They are used by mobile phones and satellite TV. Satellites can relay (pass on) microwaves from one part of the Earth to another because microwaves are not reflected or absorbed by the ionosphere. This is a layer of ions (electrically charged atoms) in the upper atmosphere. It reflects lower-frequency radio waves (see page 250). Microwaves can also be beamed across country between dish aerials on tall towers. But they do not diffract readily round obstacles because of their relatively short wavelength. So many towers are needed to give good coverage.

Mobile phones rely on a network of transmitting and receiving aerials on masts all over the country. The areas between masts are called cells.

Some clever electronics in your phone keeps switching the carrier frequency (see page 252) as you move from one cell to another. By doing this, it is possible for thousands of people to use mobile phones at the same time without overhearing other people's conversations or needing to keep their own special frequency.

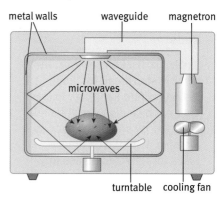

metal walls   waveguide   magnetron

microwaves

turntable   cooling fan

Microwaves produced by the magnetron pass along the waveguide and into the cooking chamber. The metal walls of the chamber reflect the microwaves inwards. The waves are also reflected by the metal grid behind the glass door because their wavelength is much larger than the holes in it. The food is rotated on a turntable to even out the effects of any 'hot spots'.

Aerials for sending and receiving radio signals to mobile phones. They are high up to provide good lines of sight to other aerials

Questions

1 *light*
  long-wavelength infrared
  short-wavelength infrared

  Which of the above:

  a can be detected by the human eye?

  b can pass through glass?

  c is (or are) emitted by red hot objects?

2 Explain why a metal bar starts to glow red when it is heated up.

3 Infrared has a heating effect on most solids. What does this tell you about whether this radiation is reflected, transmitted, or absorbed?

4 A TV remote control emits an infrared beam that carries coded information to the TV set. Explain how you might use this to find out which everyday materials absorb infrared and which transmit it (allow it to pass through).

5 List three uses of microwaves.

Radio waves

Radio waves are produced by making an electric current oscillate (making charges move to and fro) in a transmitting aerial. All radio waves have a lower frequency (and longer wavelength) than infrared. But this still covers a large range of wavelengths, so radio waves are further sub-divided into the groups shown on page 245. To carry information, radio waves must be varied in some way (see page 252). The variations, called signals, can carry sound, text, and TV pictures.

Radio waves normally travel in straight lines. As the Earth is curved, you would expect that no radio station would transmit farther than 30 or 40 miles. This is true for ground-based TV transmissions. But radio stations using the short and medium wave bands can travel much further. The reason is that the ionosphere reflects short and medium radio waves. The waves bounce between the ground and the ionosphere and so travel further from the transmitter.

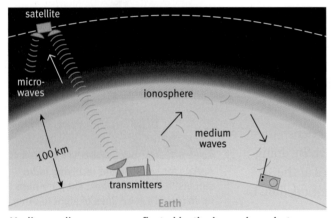

Medium radio waves are reflected by the ionosphere, but microwaves can pass through.

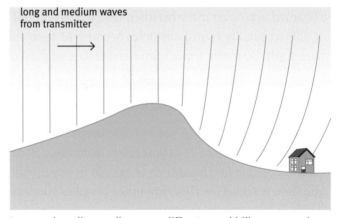

Long and medium radio waves diffract round hills, so a receiver down in a valley can still pick them up.

Diffraction also plays an important part in determining how different kinds of radio wave behave. Remember (page 239) that longer-wavelength waves diffract more readily than shorter-wavelength ones. Long and medium waves diffract (bend) around hills and other obstacles. VHF (very high frequency) is used for stereo radio and UHF (ultra-high frequency) for TV broadcasts. Because their wavelength is much shorter, these waves do not diffract so much round hills. So for good reception, there normally needs to be a straight path to your aerial. This means that more transmitting aerials are needed to give good reception everywhere.

Radio interference

If you drive along a motorway with the car radio on, you may sometimes notice that the strength of the radio signal changes. This is caused by interference between radio signals from two transmitters. As you drive along, you pass through places where the two waves add to give a stronger wave – and other places where they cancel each other out (partly) and the signal is weaker.

People who live near airports also sometimes find the radio signal varying as a plane flies overhead. This is due to interference between the radio wave reflected off the metal body of the plane and the one reaching them directly.

Bluetooth is a system that uses microwaves. It enables information to be sent between mobile phones, computers, and other devices, without any wires.

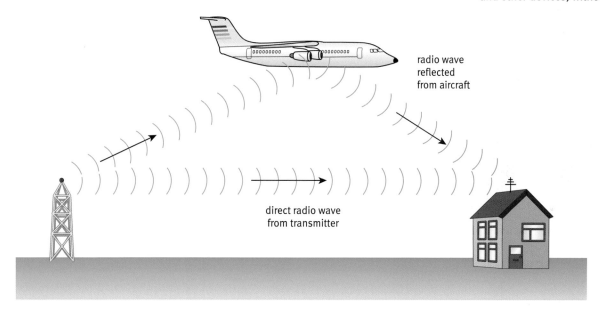

radio wave
reflected
from aircraft

direct radio wave
from transmitter

A low-flying aircraft can cause radio interference in homes nearby.

Questions

6 List two pieces of evidence that radio waves really are waves

7 A radio station is broadcasting on a frequency of 100 MHz.

   a How many radio waves does it send out every second?

   b Use the equation linking wavespeed, frequency, and wavelength to calculate the wavelength of the radio waves. (*Note*: First, you must write down the speed of radio waves in metres per second.)

8 a What is the ionosphere?

   b What effect does the ionosphere have on medium-wave (MW) radio waves? What effect does it have on microwaves?

Key words
infrared
microwaves
radio waves

Find out about:

▶ how information is carried by radio waves

How radio works

Electric charges flowing backwards, forwards, backwards, forwards . . . many times a second is called an alternating current (a.c). When there is a.c. in a piece of wire, the wire acts as an aerial. Radio waves are produced. And when radio waves reach another aerial, they generate a.c. there. This is what makes radio communication possible.

Sending sounds by radio

The diagram below shows a simple radio system. When you speak into the microphone, radio waves carry the sound from one aerial to another, and a copy of your voice comes out of the loudspeaker. The incoming sound waves might have a frequency, typically, of around 1 kHz. This is the audio frequency (AF). The radio waves have a much higher frequency, for example 1 MHz. This is the radio frequency (RF).

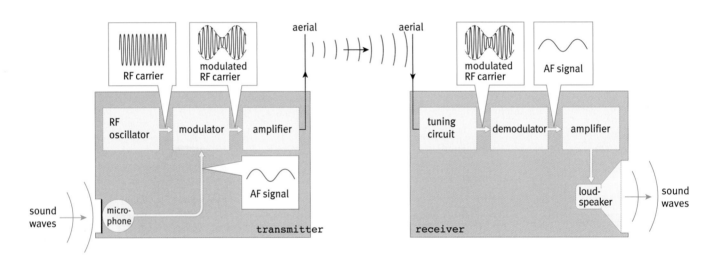

A simple radio system. When you speak into the microphone, a copy of the sound comes out of the loudspeaker.

In the microphone, the varying pressure from the sound waves produces a varying voltage – the AF signals.

The RF oscillator creates the high-frequency a.c. needed to produce the radio waves. The steady stream of waves is called the **carrier**. The AF signals are used to modulate (vary) the amplitude of the carrier so that the 'height' of the RF waves is a copy of the incoming sound waves. Before the modulated wave is sent to the aerial, its amplitude is boosted by an amplifier.

In the receiving aerial, the incoming radio waves generate electrical signals. This will be a mixture of the frequencies of all the different radio stations that are transmitting in the area. The tuning circuit selects the frequency of one RF carrier – the one carrying the radio station you want to listen to. The demodulator then removes the RF carrier, leaving only the AF signals. These are boosted by an amplifier and sent to the loudspeaker, which produces the sound.

AM and FM

The system described above uses **amplitude modulation (AM)**: the amplitude (or height) of the radio frequency (RF) carrier wave is varied by the audio frequency (AF) signals. Another method is to use **frequency modulation (FM)**, where the frequency of the carrier is varied instead of the amplitude. One advantage of FM is that the signals are less affected by electrical interference, called **noise**. Noise causes extra unwanted variations in amplitude, but with FM, this has less effect.

TV pictures are also transmitted using radio waves. A TV camera scans each scene, breaking it down into hundreds of narrow strips. Information about how the brightness and colour changes along each strip is then used to modulate the carrier.

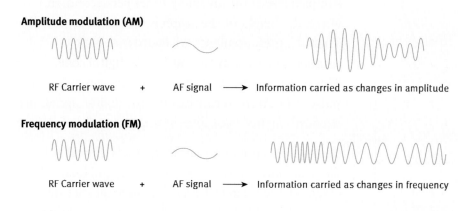

Amplitude modulation (AM)

RF Carrier wave   +   AF signal   ⟶   Information carried as changes in amplitude

Frequency modulation (FM)

RF Carrier wave   +   AF signal   ⟶   Information carried as changes in frequency

Two ways of modulating a radio frequency (RF) carrier wave, so that it carries an audio frequency (AF) signal.

Questions

1 The radio system in the diagram on the previous page has to deal with two frequencies, called RF and AF.

  a Which of these is the frequency of the sound waves?

  b Which is the frequency of the carrier?

2 Why does a radio receiver need a tuning circuit?

3 **a** In an AM radio system, what does 'modulating the carrier' mean?

  b How would this be different in an FM system?

  c What is the advantage of using FM rather than AM?

Key words

carrier
amplitude modulation (AM)
frequency modulation (FM)
noise

Going digital

Radio, TV, and telephone are all forms of telecommunication – ways of transmitting (sending) information long distances. The information may be sounds, pictures, text, or numbers, and it can be sent using wires, radio waves, or light.

Analogue and digital signals

When sound waves enter a microphone, a varying voltage is generated. The graph on the right shows how the voltage might change during one fraction of a second. Continuous variations like this are called **analogue signals**. In this case, the voltage rises and falls in the same way as the pressure in the sound waves.

The table under the graph shows how this changing voltage can be converted into **digital signals**. These are signals represented by numbers. The voltage is sampled electronically many times per second. In effect, the height of the graph is measured repeatedly. Then the measurements are coded into binary (numbers using only 0s and 1s) – a **digital code**. This is transmitted as a series of pulses (no pulse = 0; pulse = 1). At the receiving end, the digital signals are **decoded** (turned back into analogue signals).

Real systems use more sampling levels than in this diagram, much faster sampling rates, and further stages of coding in order to work more efficiently.

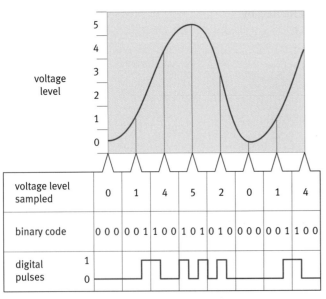

How an analogue signal can be converted into digital pulses.
(Real systems use more levels and a much faster sampling rate.)

Questions

1 A meter in which a pointer moves along a scale is called an analogue meter.

   a Why do you think it is called this?

   b What would you expect to see on a digital meter?

2 In a digital system, what does 'sampling' mean?

③ An analogue signal is converted into the following binary code:

   100 010 001 010
   100 100 010

   Using the examples in the diagram above as a guide,

   a sketch a graph showing the digital pulses

   b sketch a graph to show what the analogue signal might have looked like

Advantages of digital transmission

For transmitting information such as sounds and pictures, digital systems have several advantages over analogue ones:

- ◗ Digital signals can be handled by microprocessors (as in computers).

- ◗ Digital signals can carry more information every second than analogue ones.

- ◗ Digital signals can be delivered with no loss of quality. In other words, the sequence of 0s and 1s does not change. Analogue signals lose quality, which cannot be restored.

Here is the reason for the last point. All signals get weaker as they travel along. Noise (interference) also gets added in. But with digital signals, these effects can be corrected. Even when the incoming signals have added noise, it is still possible to tell the 0s from the 1s, as these are the only values the signal can have. It would take a very large amount of noise before it became difficult to tell the 0s and 1s apart! A regenerator can then be used to restore the pulses to their original quality. Analogue signals cannot be 'cleaned up' in this way. They can be amplified but, unfortunately, any noise is amplified as well.

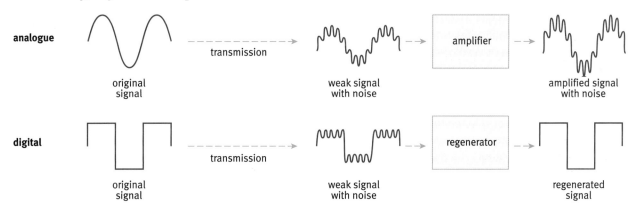

analogue — original signal — transmission — weak signal with noise — amplifier — amplified signal with noise

digital — original signal — transmission — weak signal with noise — regenerator — regenerated signal

Digital signals can be 'cleaned up' by regenerators, but when analogue signals are amplified, noise gets amplified as well.

Digital recording

A CD (compact disc) contains a metal layer with millions of tiny bumps on it, arranged in a spiral track. When you play a CD, the disc is rotated and laser light is reflected from the bumps. The reflected pulses – the light signals – are turned into electrical signals and then decoded to produce the sound. DVDs (digital versatile discs) use the same idea, although they have to store much more information in order to create full-colour TV pictures.

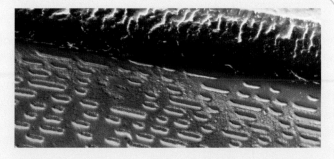

Magnified surface of a CD, showing the bumps on it (×2800)

Transmitting through optical fibres

For long-distance transmission, many telephone and cable-TV networks use optical fibres rather than radio waves, to carry the signal (see page 242). These long, thin strands of glass or clear plastic carry digital signals in the form of very brief pulses of light (or infrared). In effect, a light beam is being switched on and off very rapidly to produce a digital code of 1s and 0s. At the transmitting end, electrical signals are changed into light signals by an LED (light-emitting diode) or a diode-laser. At the receiving end, the light signals are picked up by a photodiode and changed back into electrical signals.

Optical fibre cables are thinner and lighter than electric cables. They can also carry many more signals. With electrical cables, the signal gets weaker quite quickly, so amplifiers are needed at regular intervals to boost it again. With optical fibres, these 'booster stations' can be much farther apart. Optical fibres are not affected by electrical interference and they cannot be 'tapped' (people cannot listen in to others' messages).

Optical fibres from inside a cable. Most telephone calls between Britain and America are carried by a cable like this, lying along the bottom of the Atlantic Ocean.

Digital radio

DAB (Digital Audio Broadcasting) can handle more stations than the older analogue AM and FM systems and give high-quality, interference-free reception. With the older systems, reflections from hills and buildings cause interference. DAB actually uses these 'multipath' signals to improve quality! Its transmitters send out radio waves in pulses. In the receiver, a microchip does some clever processing on the various incoming signals, and uses them to produce a stronger signal.

DAB transmitters all use the same basic carrier frequency. However, this does not mean that you can only listen to one station. Digital signals from up to ten stations are split into groups in such a way that they can be transmitted alongside each other. The receiver unscrambles the different groups when they arrive. This system is called multiplexing. It is also used by telephone networks so that lots of calls can be sent together along the same cable.

A digital radio (DAB) receiver

Key words

analogue signals
digital signals
digital code
decoded

Questions

4 Why is it possible to remove noise from digital signals but not from analogue ones?

5 List as many reasons as you can why optical fibres are better than electric cables for long-distance communication.

K Radiation from space

Getting through

Only some types of electromagnetic wave can pass through the Earth's atmosphere. The graph below shows the percentage of radiation of each frequency reaching the ground. Radiation of many frequencies is completely absorbed by the atmosphere.

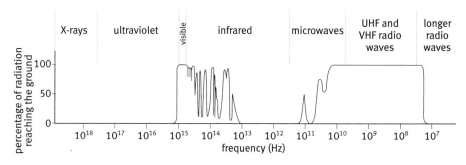

Graph showing the percentage of radiation reaching the earth's surface across the electromagnetic spectrum

One of the two main bands that can pass through the atmosphere is visible light (the band of frequencies that our eyes can detect.)

The only other large band of radiations that get through the atmosphere is lower-frequency microwaves and UHF and VHF radio waves. This is why radio telescopes have been developed.

Listening for aliens

As well as observing natural radio emissions from outer space, radio telescopes can be used to search for extraterrestrial intelligence. We on Earth have to hope that intelligent life is capable of sending out radio signals. The search for these signals is called the SETI project (the Search for ExtraTerrestrial Intelligence). It has been going on for over 40 years. If we ever do detect signals that we think come from extraterrestrials, having conversations with them would be almost impossible because of the time delay. Our radio signals would take nearly four years to reach even the nearest star. Then we would have to wait another four years for a reply.

Nowadays astronomers also use telescopes that detect infrared and ultraviolet, but the telescope itself has to be in space, outside the Earth's atmosphere. Information collected is then relayed back to Earth using radio signals which can pass through the atmosphere.

Find out about:

▶ which electomagnetic radiation can get through the atmosphere
▶ how astronomers make use of this

The big bang, coming to a screen near you!

Scientists think that the Universe began about 14 billion years ago with a huge explosion called the big bang. Radiation from the big bang is still reaching us from every direction in space – it is called the **microwave background**. If you turn on a TV and look at a channel that hasn't been tuned in, you will probably see lots of dancing spots on the screen. Much of this 'snow' is due to the microwave background. You are seeing the remnants of the big bang!

Radio telescopes like this one are mainly used to detect the natural radio waves coming from distant stars and galaxies. However, they can also be used to search for signals from intelligent extraterrestrials.

Key words
microwave background

Summary

This module has introduced you to some basic ideas about waves, and how they behave. Light and other members of the electromagnetic spectrum behave like waves. Several types of electromagnetic wave (including radio, microwaves, infrared, and visible light) are used nowadays for communications.

Waves

▹ A wave is a disturbance moving through a medium. Parts of the medium move to and fro as the wave passes, but the medium does not move as a whole in the direction of the wave.

▹ A wave carries energy and information through the medium.

Wave characteristics

▹ The amplitude of a wave is the maximum disturbance of each particle of the medium as the wave passes.

▹ The frequency is the number of waves produced every second by the source (which is also the number of waves passing any point in the medium every second).

▹ The wave speed is the speed at which each wave crest moves through the medium.

▹ Amplitude and frequency depend on the source; wave speed depends on the medium.

▹ The wavelength is the distance from one wave crest to the next.

▹ For all waves:
'wave speed = frequency × wavelength'.
In any given medium, the wave speed is fixed. So the bigger the frequency, the smaller the wavelength.

Wave properties

▹ Four ways in which all waves behave are: reflection, refraction, diffraction, and interference.

▹ The last two are characteristic of waves. They provide evidence that a radiation behaves like a wave.

Light waves

▹ Light behaves like a wave, showing diffraction and interference, if you look carefully.

▹ Scientists believe that light consists of vibrating electric and magnetic fields – an electromagnetic wave.

Electromagnetic spectrum

▹ The electromagnetic 'family' of waves all travel through vacuum with a wave speed of 300 000 km per second.

▹ Electromagnetic waves have very different properties, depending on their frequency. In order of increasing frequency, they are: radio, microwaves, infrared, visible light, ultraviolet, X-rays, gamma rays.

▹ The three highest-frequency types are ionizing: they can cause chemical changes in materials that absorb them.

Communications

▹ Electromagnetic waves can be used to carry information. This is 'coded' on to a carrier wave, as changes in amplitude or frequency (analogue signals), or by pulsing the beam on and off very rapidly (digital signals).

▹ Digital signals can be communicated more accurately, with less unwanted 'noise'. H

Questions

1 Copy this table and complete it by writing either *gets bigger*, *gets smaller*, or *stays the same* in each cell:

		amplitude	frequency	wavelength	wave speed
a	waves on the surface of a pond, as they travel away from the disturbance that caused them				
b	a wave on a long spring, when the end is moved up and down more rapidly				
c	water waves being diffracted as they pass through a gap in a barrier				
d	a light wave as it moves from air into a clear plastic block				
e	radio waves as they travel from a satellite to a receiving dish on the ground				

2 Sketch diagrams to show what happens to plane water waves in each of the following situations:

   a hitting a straight barrier, at an angle of 45°

   b moving from shallow into deeper water

   c hitting an obstacle whose width is about 4 wavelengths

3 Describe what you would see as two sets of water waves, of exactly the same frequency and travelling across the same water surface at right angles to each other, met and crossed. Sketch a diagram if it helps you explain. Which wave property is involved here?

4 Make a list of all the pieces of evidence you would use to convince someone that the radiations of the electromagnetic spectrum behave like waves.

5 Make a table with four columns. In the left-hand column, list the following types of electromagnetic radiation:

   radio waves, microwaves, infrared, light, ultraviolet, X-rays, gamma rays.

   Complete the other three columns to show, for each type of radiation:

   main sources, ways of detecting, main uses.

6 All electromagnetic waves travel at a speed of 300 000 km/s in a vacuum (and at much the same speed in air).

   a How long does it take for a radio signal to travel from Britain to New Zealand (18 000 km)?

   b How long does it take for a radio message to travel to Earth from an astronaut on the Moon (320 000 km)?

   c How long does it take for light from the Sun to reach Earth (150 000 000 km)?

   d When the *Voyager 2* probe passed Neptune, it was approximately 4 500 000 000 km from Earth. It sent back the photos it took as radio signals. How long did they take to reach us?

   e The nearest star (after the Sun) is 4.35 light-years away, i.e. the distance light travels in 4.35 years. How many kilometres is this?

7 Optical fibres are increasingly important for communications. Explain why a light ray travels along an optical fibre, following any bends in it, and not escaping through the sides. Include a diagram to show what is happening.

Glossary

absorbed The radiation which hits an object and is not reflected, or transmitted through it, is absorbed. The object gets a little hotter as a result.

abundant Abundance measures how common an element is. Silicon is abundant in the lithosphere. Nitrogen is abundant in the atmosphere.

acid A compound that dissolves in water to give a solution with a pH lower than 7. Acid solutions change the colour of indicators, form salts when they neutralize alkalis, react with carbonates to form carbon dioxide, and give off hydrogen when they react with a metal. An acid is a compound that contains hydrogen in its formula and produces hydrogen ions when it dissolves in water.

active site The part of an enzyme where the reacting molecules fit into.

active transport Molecules are moved in or out of a cell using energy. This process is used when transport needs to be faster than diffusion, and when molecules are being moved from a region where they are at low concentration to where they are at high concentration.

active 'working memory' One explanation for how the human memory works.

actual yield The mass of the required chemical obtained after separating and purifying the product of a chemical reaction.

ADH A hormone making tubules more permeable to water, causing greater re-absorption of water.

aerial A wire, or arrangement of wires, that emits radio waves when there is an alternating current in it, and in which an alternating current is induced by passing radio waves. So it acts as a source or a receiver of radio waves.

alkali A compound that dissolves in water to give a solution with a pH higher than 7. An alkali can be neutralized by an acid to form a salt. Solutions of alkalis contain hydroxide ions.

alternating current (a.c.) An electric current that reverses direction many times a second.

amino acids The small molecules which are joined in long chains to make proteins. All the proteins in living things are made from 20 different amino acids joined in different orders.

ammeter A meter that measures the size of an electric current in a circuit.

ampere (or amp, for short) The unit of electric current.

amplifier A device for increasing the amplitude of an electrical signal. Used in radios and other audio equipment.

amplitude For a mechanical wave, the maximum distance that each point on the medium moves from its normal position as the wave passes. For an electromagnetic wave, the maximum value of the varying electric field (or magnetic field).

amplitude modulation (AM) One way in which a radio wave can be made to carry an audio signal. The amplitude of the carrier wave is made to vary in the same way as the sound wave.

analogue signals Signals used in communications in which the amplitude can vary continuously.

antagonistic effectors Antagonistic effectors have opposite effects.

antibodies A group of proteins made by white blood cells to fight dangerous microorganisms. A different antibody is needed to fight each different type of microorganism. Antibodies bind to the surface of the microorganism, which triggers other white blood cells to digest them.

asexual reproduction When a new individual is produced from just one parent.

atmosphere The layer of gases that surrounds the Earth.

attract Pull towards.

attractive forces (between molecules) Forces that try to pull molecules together. Attractions between molecules are weak. Molecular chemicals have low melting points and boiling points because the molecules are easy to separate.

auxins A plant hormone that affects plant growth and development. For example, auxin stimulates growth of roots in cuttings.

average speed The distance moved by an object divided by the time taken for this to happen.

axon A long, thin extension of the cytoplasm of a neuron. The axon carries electrical impulses very quickly.

balanced chemical equation An equation showing the formulae of the reactants and products. The equation is balanced when there is the same number of each kind of atom on both sides of the equation.

base pairing The bases in a DNA molecule (A, C, G, T) always bond in the same way. A and T always bond together. C and G always bond together.

behaviour Everything an organism does; its response to all the stimuli around it.

biosphere All the living organisms on Earth. This includes all the plants, animals, and microorganisms.

bleach A chemical that can destroy unwanted colours. Bleaches also kill bacteria. A common bleach is a solution of chlorine in sodium hydroxide.

bulk chemicals Chemicals made by industry on a scale of thousands or millions of tonnes per year. Sulfuric acid, sodium hydroxide, and chlorine are examples.

burette A graduated tube with a tap or valve. A burette is used to measure the volume of a solution added during a titration.

carbohydrate A natural chemical made of carbon, hydrogen, and oxygen. The hydrogen and oxygen are present in the same proportions as in water. An example is glucose, $C_6H_{12}O_6$. Carbohydrates includes sugars, starch, and cellulose.

carbon cycle The cycling of the element carbon in the environment between the atmosphere, biosphere, hydrosphere, and lithosphere. The element exists in different compounds in these spheres. In the atmosphere it is mainly present as carbon dioxide.

carbonate A compound which contains carbonate ions, CO_3^{2-}. An example is calcium carbonate, $CaCO_3$.

carrier A steady stream of radio waves produced by an RF oscillator in a radio.

catalyst A chemical which speeds up a chemical reaction but is not used up in the process.

central nervous system In mammals the brain and spinal cord.

cerebral cortex The highly folded outer region of the brain, concerned with conscious behaviour.

charged Carrying an electric charge. Some objects (such as electrons, protons) are permanently charged. A plastic object can be charged by rubbing it. This transfers electrons to or from it.

chemical change A reaction that forms new chemicals.

chemical equation A summary of a chemical reaction showing the reactants and products with their physical states (*see* balanced chemical equation).

chemical properties A chemical property describes how an element or compound interacts with other chemicals, for example the reactivity of a metal with water.

chromosomes Long, thin, thread-like structures in the nucleus of a cell made from a molecule of DNA. Chromosomes carry the genes.

clone A new cell or individual made by asexual reproduction. A clone has the same genes as its parent.

coding In communications, converting an analogue signal into a digital one.

collision theory The theory that reactions happen when molecules collide. The theory helps to explain the factors that affect the rates of chemical change. Not all collisions between molecules lead to reaction.

concentration The quantity of a chemical dissolved in a stated volume of solution. Concentrations can be measured in grams per litre.

consciousness The part of the human brain concerned with thought and decision making.

conservation of energy The principle that the total amount of energy at the end of any process is always equal to the total amount of energy at the beginning – though it may now be stored in different ways and in different places.

core (of the body) Central parts of the body where the body temperature is kept constant.

corrosive A corrosive chemical may destroy living tissue on contact.

counter-force A force in the opposite direction to something's motion.

covalent bonding Strong attractive forces that hold atoms together in molecules. Covalent bonds form between atoms of non-metallic elements.

crystal A piece of a solid with a giant covalent or ionic structure and a regular shape. Crystals are shiny and have faces at a particular angle to each other. This is caused by the regular arrangement of atoms or ions in the giant structure.

crust (of the Earth) The outer layer of the lithosphere

cuttings A shoot or leaf taken from a plant, to be grown into a new plant.

decoding In communications, converting a digital signal back into an analogue one.

development How an organism changes as it grows and matures. As a zygote develops, it forms more and more cells. These are organised into different tissues and organs.

diffraction What happens when waves hit the edge of a barrier or pass through a gap in a barrier. They bend a little and spread into the region behind the barrier.

diffusion The passive movement of molecules from areas where they are more concentrated to areas where they are less concentrated.

digital code A series of 0s and 1s that can be used to represent an analogue signal, and from which that signal can be reconstructed.

digital signals Signals used in communications in which the amplitude can take only one of two values, corresponding to the digits 0 and 1.

direct current (d.c.) An electric current that stays in the same direction.

dispersion The splitting of white light into different colours (frequencies), for example by a prism.

dissolve Some chemicals dissolve in liquids (solvents). Salt and sugar, for example, dissolve in water.

distance–time graph A useful way of summarizing the motion of an object by showing how far it has moved from its starting point at every instant during its journey.

DNA The chemical that makes up chromosomes – deoxyribonucleic acid. DNA carries the genetic code, which controls how an organism develops.

double helix The shape of the DNA molecule, with two strands twisted together in a spiral.

driving force The force pushing something forward, for example a bicycle.

Ecstasy A recreational drug that increases the concentration of serotonin at the synapses in the brain, giving pleasurable feelings. Long-term effects may include destruction of the synapses.

effector The part of a control system that brings about a change to the system.

efficiency The percentage of the energy supplied to a device that is transferred to the place we want, or in the way we want. For example, to find the efficiency of a kettle we would divide the gain in energy of the water by the work done on the kettle element by the electricity supply, and multiply by 100.

electric charge A fundamental property of matter. Electrons and protons are charged particles. Objects become charged when electrons are transferred to or from them, for example by rubbing.

electric circuit A closed loop of conductors connected between the positive and negative terminals of a battery or power supply.

electric current A flow of charges around an electric circuit.

electric field A region where an electric charge experiences a force. There is an electric field around any electric charge.

electrode A conductor made of a metal or graphite through which a current enters or leaves a chemical during electrolysis. Electrons flow into the negative electrode (cathode) and out of the positive electrode (anode).

electrolysis Splitting up a chemical into its elements by passing an electric current through it.

electrolyte A chemical which can be split up by an electric current when molten or in solution is the electrolyte. Ionic compounds are electrolytes.

electromagnetic induction The name of the process in which a potential difference (and hence often an electric current) is generated in a wire, when it is in a changing magnetic field.

electromagnetic spectrum The 'family' of electromagnetic waves of different frequencies and wavelengths.

electromagnetic wave A wave consisting of vibrating electric and magnetic fields, which can travel in a vacuum. Visible light is one example.

electron A tiny particle which is part of an atom. Electrons are found outside the nucleus. Electrons have negligible mass and are negatively charged, 1−.

electron configuration The number and arrangement of electrons in an atom of an element.

electrostatic attraction The force of attraction between objects with opposite electric charges. A positive ion, for example, attracts a negative ion.

embryo The earliest stage of development for an animal or plant. In humans the embryo stage lasts until age two months.

end point The point during a titration at which the reaction is just complete. For example, in an acid–alkali titration, the end-point is reached when the indicator changes colour. This happens when exactly the right amount of acid has been added to react with all the alkali present at the start.

energy level The electrons in an atom have different energies and are arranged at distinct energy levels.

enzyme A protein that catalyses (speeds up) chemical reactions in living things.

excretion The removal of waste products of chemical reactions in cells.

extraction (of metals) The process of obtaining a metal from a mineral by chemical reduction or electrolysis. It is often necessary to concentrate the ore before extracting the metal.

extremities Parts of the body away from the core, for example fingers.

fatty sheath Fat wrapped around the outside of an axon to insulate neurons from each other.

feral A child who has not learned to speak because they have not been exposed to language early enough in life.

fine chemicals Chemicals made by industry in smaller quantities than bulk chemicals. Fine chemicals are used in products such as food additives, medicines, and pesticides.

flame colour A colour produced when a chemical is held in a flame. Some elements and their compounds give characteristic colours. Sodium and sodium compounds, for example, give bright yellow flames.

force A push or a pull experienced by an object when it interacts with another. A force is needed to change the motion of an object.

formula (chemical) A way of describing a chemical that uses symbols for atoms. A formula gives information about the numbers of different types of atom in the chemical. The formula of sulfuric acid, for example, is H_2SO_4.

frequency The number of waves produced every second by a source, or that pass a point in the medium every second.

frequency modulation (FM) One way in which a radio wave can be made to carry an audio signal. The frequency of the carrier wave is made to vary in the same way as the sound wave.

friction The force exerted on an object due to the interaction between it and another object that it is sliding over, or tending to slide over. It is caused by the roughness of both surfaces at a microscopic level.

gametes The sex cells that fuse to form a zygote. In humans, the male gamete is the sperm and the female gamete is the egg.

gamma rays Very high frequency electromagnetic waves emitted by some radioactive materials.

gemstone A crystalline rock or mineral that can be cut and polished for jewellery.

genes A section of DNA giving the instructions for a cell about how to make one kind of protein.

gene switching Genes in the nucleus of a cell switch off and are inactive when a cell becomes specialized. Only genes which the cell needs to carry out its particular job stay active.

genetic variation Differences between individuals caused by differences in their genes. Gametes show genetic variation – they all have different genes.

giant covalent structure A giant, three-dimensional arrangement of atoms that are held together by covalent bonds. Silicon dioxide and diamond have giant covalent structures.

giant ionic structure The structure of solid ionic compounds. There are no individual molecules, but millions of oppositely charged ions packed closely together in a regular, three-dimensional arrangement.

glands Parts of the body that make enzymes, hormones, and other secretions in the body, for example sweat glands.

granite A hard igneous rock with clearly visible crystals of various minerals.

gravitational potential energy The energy that something has owing to its position above the ground.

group Each column in the periodic table is a group of similar elements.

halogens The family name of the group 7 elements.

heat stroke A life-threatening rise in body temperature where the body temperature control system fails.

homeostasis Keeping a steady state inside your body.

hydrogen ion A hydrogen atom that has lost one electron. The symbol for a hydrogen ion is H^+. Acids produce aqueous hydrogen ions, $H^+(aq)$, when dissolved in water.

hydrosphere All the water on Earth. This includes oceans, lakes, rivers, underground reservoirs, and rainwater.

hydroxide ion A negative ion, OH^-. Alkalis give aqueous hydroxide ions when they dissolve in water.

hypothalamus The part of the brain that controls many different functions, for example body temperature.

hypothermia A fall in body temperature to below 35 °C.

in parallel A way of connecting electric components that makes a branch (or branches) in the circuit so that charges can flow round more than one loop.

in series A way of connecting electric components so that all in a single loop. The charges pass through them all in turn.

indicator A chemical that shows whether a solution is acidic or alkaline. For example, litmus turns blue in alkalis and red in acids. Universal indicator has a range of colours that show the pH of a solution.

infrared Electromagnetic waves with a frequency lower than that of visible light, beyond the red end of the visible spectrum.

instantaneous speed The speed of an object at a particular instant. In practice, its average speed over a very short time interval.

intensity The amount of energy transferred every second by a beam of radiation. The brightness of a light beam indicates its intensity.

interaction What happens when two objects collide, or influence each other at a distance. When two objects interact, each experiences a force.

interaction pair Two forces that arise from the same interaction. They are equal in size and opposite in direction, and each acts onn a different object.

interference What happens when two waves meet. If the waves have the same frequency, an interference pattern is formed. In some places, crests add to crests, forming bigger crests; in other places, crests and troughs cancel each other out.

involuntary An automatic response made by the body without conscious thought.

ionic bonding Very strong attractive forces that hold the ions together in an ionic compound. The forces come from the attraction between positively and negatively charged ions.

ionic compounds Compounds formed by the combination of a metal and non-metal. They contain positively charged metal ions and negatively charged non-metal ions.

ionizing radiation Radiation that can remove electrons from atoms in its path. Ionizing radiation, such as ultraviolet, X-rays, and gamma rays, can damage living cells.

ions An electrically charged atom or group of atoms.

kidneys Organs in the body which remove waste urea from the blood, and balance water and salt levels.

kinetic energy The energy which something has owing to its motion.

learned behaviour Responses to stimuli which an organism develops because of experience.

light-dependent resistor (LDR) An electric circuit component whose resistance varies depending on the brightness of light falling on it.

line spectrum A spectrum made up of a series of lines. Each element has its own characteristic line spectrum.

lithosphere The rigid outer layer of the Earth, made up of the crust and the part of mantle just below it.

longitudinal wave A wave in which the particles of the medium vibrate in the same direction as the wave is travelling. Sound is an example.

long-term memory The part of the memory that stores information for a long period, or permanently.

mantle The layer of rock between the crust and the outer core of the Earth. It is approximately 2900 km thick.

medium A material through which a wave travels. The plural is 'media'.

meiosis Cell division that halves the number of chromosomes to produce gametes. The four new cells are genetically different from each other and from the parent cell.

memory The storage and retrieval of information by the brain.

meristem cells Unspecialized cells in plants that can develop into any kind of specialized cell.

metal Elements on the left side of the periodic table. Metals have characteristic properties: they are shiny when polished and conduct electricity. Some metals react with acids to give salts and hydrogen. Metals are present as positive ions in salts.

metal hydroxide A compound consisting of metal positive ions and hydroxide ions. Examples are sodium hydroxide, NaOH, and magnesium hydroxide, $Mg(OH)_2$.

metal oxide A compound of a metal with oxygen.

metallic bonding Very strong attractive forces that hold metal atoms together in a solid metal. The metal atoms lose their outer electrons and form positive ions. The electrons drift freely around the lattice of positive metal ions and hold the ions together.

microwaves Radio waves of the highest frequency (shortest wavelength), used for mobile phones and satellite TV.

mineral A naturally occurring element or compound in the Earth's lithosphere.

mitochondria An organelle in plant cells where respiration takes place.

mitosis Cell division that makes two new cells identical to each other and to the parent cell.

models of memory Explanations for how memory is structured in the brain.

modulate To vary the amplitude or frequency of the carrier waves produced in a radio so that they carry the information in a sound wave.

molecular models Models to show the arrangement of atoms in molecules, and the bonds between the atoms.

molecule A group of atoms joined together. Most non-metals consist of molecules. Most compounds of non-metals with other non-metals are also molecular.

molten A chemical in the liquid state. A chemical is molten when the temperature is above is melting point but below its boiling point.

momentum A property of any moving object. Equal to mass multiplied by velocity. The plural is momenta.

motor neuron A neuron that carries nerve impulses from the brain or spinal cord to an effector.

mRNA Messenger RNA, a chemical involved in making proteins in cells. The mRNA molecule is similar to DNA but single stranded. It carries the genetic code from the DNA molecule out of the nucleus into the cytoplasm.

multi-store model One explanation for how the human memory works.

muscles Muscles move parts of the skeleton for movement. There is also muscle tissue in other parts of the body, for example in the walls of arteries.

nanometres A unit used for microscopic measurements. 1 nm = 0.001 µm = 0.000 001 mm

natural cycle (in the environment) The cycling of an element between the atmosphere, hydrosphere, and biosphere as a result of natural processes.

negative A label used to name one type of charge, or one terminal of a battery. It is the opposite of positive.

negative feedback When a change in a system results in an action that will reverse the change, bringing the system back to its normal state.

negative ion An ion that has a negative charge (an anion).

nerve impulses Electrical signals carried by neurons (nerve cells).

nervous system Tissues and organs which control the body's responses to stimuli. In a mammal it is made up of the central nervous system and peripheral nervous system.

neuroscientist A scientist who studies how the brain and nerves function.

neutralization A reaction in which an acid reacts with an alkali to form a salt. During neutralization reactions, the hydrogen ions in the acid solution react with hydroxide ions in the alkaline solution to make water molecules.

neutrons An uncharged particle found in the nucleus of atoms. The relative mass of a neutron is 1.

newborn reflexes Reflexes to particular stimuli that usually occur only for a short time in newborn babies.

nitrogen cycle The continual cycling of nitrogen, which is one of the elements that is essential for life. By being converted to different chemical forms, nitrogen is able to cycle between the atmosphere, lithosphere, hydrosphere, and biosphere.

noise Unwanted electrical signals that get added on to radio waves during transmission, causing additional modulation. Sometimes called 'interference'.

nucleus The tiny central part of an atom made up of protons and neutrons. Most of the mass of an atom is concentrated in its nucleus.

Ohm's law The result that the current, I, through a resistor, R, is proportional to the voltage, V, across the resistor, provided its temperature remains the same. Ohm's law does not apply to all conductors.

one-gene–one-protein theory The idea that each gene on a chromosome controls the production of one protein in the cell.

optical fibres Thin glass fibres, down which a light beam can travel. The beam is reflected at the sides by total internal reflection, so very little escapes. Used in modern communications, for example, to replace wires for carrying telephone signal.

optimum temperature The temperature at which enzymes work fastest.

ore A natural mineral that contains enough valuable minerals to make it profitable to mine.

organelles The specialized parts of a cell, such as the nucleus and mitochondria. Chloroplasts are organelles that occur only in plant cells.

organs Parts of a plant or animal made up of different tissues.

osmosis The diffusion of water across a partially permeable membrane.

oxidation A reaction that adds oxygen to a chemical.

oxide A compound of an element with oxygen.

partially permeable membrane A membrane that acts as a barrier to some molecules but allows others to diffuse through freely.

percentage yield A measure of the efficiency of a chemical synthesis.

$$\text{percentage yield} = \frac{\text{actual yield}}{\text{theoretical yield}}$$

period In the context of chemistry, a row in the periodic table.

periodic In chemistry, a repeating pattern in the properties of elements. In the periodic table one pattern is that each period starts with metals on the left and ends with non-metals on the right.

peripheral nervous system The network of nerves connecting the central nervous system to the rest of the body.

pH scale A number scale that shows the acidity or alkalinity of a solution in water.

phloem A plant tissue which transports sugar throughout a plant.

photons Tiny 'packets' of electromagnetic radiation. All electromagnetic waves are emitted and absorbed as photons. The energy of a photon is proportional to the frequency of the radiation.

photosynthesis The process in green plants which uses energy from sunlight to convert carbon dioxide and water into the sugar glucose.

phototropism The bending of growing plant shoots towards the light.

physical properties Properties of elements and compounds such as melting point, density, and electrical conductivity. These are properties that do not involve one chemical turning into another.

pituitary gland Part of the human brain which coordinates many different functions, for example release of ADH.

positive A label used to name one type of charge, or one terminal of a battery. It is the opposite of negative.

positive ions Ions that have a positive charge (cations).

potential difference (p.d.) The difference in potential energy (for each unit of charge flowing) between any two points in an electric circuit.

power In an electric circuit, the rate at which work is done by the battery or power supply on the components in a circuit. Power is equal to current × voltage.

processing centre The part of a control system that receives and processes information from the receptor, and triggers action by the effectors.

protein Chemicals in living things that are polymers made by joining together amino acids.

proton number The number of protons in the nucleus of an atom (also called the atomic number). In an uncharged atom this also gives the number of electrons.

protons Tiny particles that are present in the nuclei of atoms. Protons are positively charged, 1+.

Prozac A brand name for an antidepressant drug. It increases the concentration of serotonin at the synapses in the brain.

quartz A crystalline form of silicon dioxide, SiO_2.

radiation A flow of information and energy from a source. Light and infrared are examples. Radiation spreads out from its source, and may be absorbed or reflected by objects in its path. It may also go (be transmitted) through them.

radio waves Electromagnetic waves of a much lower frequency than visible light. They can be made to carry signals and are widely used for communications.

rate of reaction A measure of how quickly a reaction happens. Rates can be measured by following the disappearance of a reactant or the formation of a product.

reactants The chemicals on the left-hand side of an equation. These chemicals react to form the products.

Reacting mass The masses of chemicals that react together, and the masses of products that are formed. Reacting masses are calculated from the balanced symbol equation using relative atomic masses and relative formula masses.

reaction (of a surface) The force exerted by a hard surface on an object that presses on it.

reactive metal A metal with a strong tendency to react with chemicals such as oxygen, water, and acids. The more reactive a metal, the more strongly it joins with other elements such as oxygen. So reactive metals are hard to extract from their ores.

receptor The part of a control system that detects changes in the system and passes this information to the processing centre.

reducing agent A chemical that removes oxygen from another chemical. For example, carbon acts as a reducing agent when it removes oxygen from a metal oxide. The carbon is oxidized to carbon monoxide during this process.

reduction A reaction that removes oxygen from a chemical. For example, some metal oxides can be reduced to metals by heating them with carbon.

reflection What happen when a wave hits a barrier and bounces back off it. If you draw a line at right angles to the barrier, the reflected wave has the same angle to this line as the incoming wave.

reflex arc A neuron pathway that brings about a reflex response. A reflex arc involves a sensory neuron, connecting neurons in the brain or spinal cord, and a motor neuron.

refraction Waves change their wavelength if they travel from one medium to another in which their speed is different. For example, when travelling into shallower water, waves have a smaller wavelength as they slow down.

relative atomic mass The mass of an atom of an element compared to the mass of an atom of carbon. The relative atomic mass of carbon has been defined as 12.

relative formula mass The combined relative atomic masses of all mass the atoms in a formula. To find the relative formula mass of a chemical, you just add up the relative atomic masses of the atoms in the formula.

repel Push apart.

reproductive cloning Making new individuals that are genetically identical to the parent.

resistance The resistance of a component in an electric circuit indicates how easy or difficult it is to push charges through it.

respiration A series of chemical reactions in cells which release energy for the cell to use.

response Action or behaviour that is caused by a stimulus.

resultant force The sum, taking their directions into account, of all the forces acting on an object.

ribosomes Organelles in cells. Amino acids are joined together to form proteins in the ribosomes.

risk assessment A check on the hazards involved in a scientific procedure. A full assessment include the steps to be taken to avoid or reduce the risks from the hazards identified.

rooting powder A product used in gardening containing plant hormones. Rooting powder encourages a cutting to form roots.

salt An ionic compound formed when a metal reacts with a non-metal or when an acid neutralizes an alkali.

sampling In the context of physics, measuring the amplitude of an analogue signal many times a second in order to convert it into a digital signal.

sandstone A rock made of sand grains stuck together.

scale up To redesign a synthesis to produce a chemical in larger amounts. A process might be scaled up first from a laboratory method to a pilot plant; then from a pilot plant to a full-scale industrial process.

sensory neuron A neuron that carries nerve impulses from a receptor to the brain or spinal cord.

serotonin A chemical released at one type of synapse in the brain, resulting in feelings of pleasure.

shell A region in space (around the nucleus of an atom) where there can be electrons.

shivering Very quick muscle contractions. Releases more energy from muscle cells to raise body temperature.

short-term memory The part of the memory that stores information for a short time.

signal An electromagnetic wave with variations in its amplitude or frequency, or being rapidly switched on an off, which is used to carry information.

simple reflex An automatic response made by an animal to a stimulus.

small molecules Particles of chemicals that consist of small numbers of atoms bonded together. Chemicals made up of one or more non-metallic element and which have low boiling and melting points consist of small molecules.

specialized A specialized cell is adapted for a particular job.

spectroscopy The use of instruments to produce and analyse spectra. Chemists use spectroscopy to study the composition, structure, and bonding of elements and compounds.

spectrum One example is the continuous band of colours, from violet to red, produced by shining white light through a prism. Passing light from a flame test through a prism produces a line spectrum.

speed of light 300 000 kilometres per second – the speed of all electromagnetic waves in a vacuum.

static electricity Electric charge that is not moving round a circuit but has built up on an object such as a comb or a rubbed balloon.

stem cells Unspecializzed animal cells that can divide and develop into specialized cells.

stimulus A change in the environment that causes a response.

structural proteins Proteins which are used to build cells.

surface area (of a solid chemical) The area of a solid in contact with other reactants that are liquids or gases.

synapses A tiny gap between neurons that transmits nerve impulses from one neuron to another by means of chemicals diffusing across the gap.

theoretical yield The amount of product that would be obtained in a reaction if all the reactants were converted to products exactly as described by the balanced chemical equation.

therapeutic cloning Growing new tissues and organs from cloning embryonic stem cells. The new tissues and organs are used to treat people who are ill or injured.

thermistor An electric circuit component whose resistance changes markedly with its temperature. It can therefore be used to measure temperatures.

tissue Group of specialized cells of the same type working together to do a particular job.

titration An analytical technique used to find the exact volumes of solutions that react with each other.

total internal reflection (TIR) What can happen when light hits a boundary with a medium where it moves faster (for example, going from glass into air). If the light hits the boundary at an angle greater than the critical angle, it is reflected.

toxic A chemical which may lead to serious health risks, or even death, if breathed in, swallowed, or taken in through the skin.

transformer An electrical device, consisting of two coils of wire wound on an iron core. An alternating current in one coil causes an ever-changing magnetic field which induces an alternating current in the other. Used to 'step' voltage up or down to the level required.

transmitted (transmit) When radiation hits an object, it may go through it. It is said to be transmitted through it. We also say that a radio aerial transmits a signal. In this case, transmit means 'emits' or 'sends out'.

transverse wave A wave in which the particles of the medium vibrate at right angles to the direction in which the wave is travelling. Water waves are an example.

trends A description of the way a property increases or decreases along a series of elements or compounds which is often applied to the elements (or their compounds) in a group or period.

triplet code A sequence of three bases coding for a particular amino acid in the genetic code.

ultraviolet Electromagnetic waves with frequencies higher than those of visible light, beyond the violet end of the visible spectrum.

unspecialized Cells which have not yet developed into one particular type of cell.

urea A waste product made by the liver from the breakdown of amino acids the body does not use.

urine Waste excreted from the body in the kidneys and stored in the bladder.

vasoconstriction Narrowing of blood vessels.

vasodilation Widening of blood vessels.

velocity The speed of an object in a given direction. Unlike speed, which only has a size, velocity also has a direction.

velocity-time graph A useful way of summarizing the motion of an object by showing its velocity at every instant during its journey.

vibrates Moves rapidly and repeatedly back and forth.

voltage The voltage marked on a battery or power supply is a measure of the 'push' it exerts on charges in an electric circuit. The 'voltage' between two points in a circuit means the 'potential difference' between these points.

wave speed The speed at which waves move through a medium.

wavelength The distance between one wave crest (or wave trough) and the next.

weathering Chemical changes of the minerals in rocks caused by reactions with air and water.

work Work is done whenever a force makes something move. The amount of work is force multiplied by distance moved in the direction of the force. This is equal to the amount of energy transferred.

X-rays Electromagnetic waves with high frequency, well above that of visible light.

xylem A plant tissue which transports water through a plant.

zygote The cell made when a sperm cell fertilizes an egg cell in sexual reproduction.

Index

Publisher's acknowledgements

Oxford University Press wishes to thank the following for their kind permission to reproduce copyright material:

p10 Dimitri Iundt/Corbis UK Ltd.; **p11 t** Publiphoto Diffusion/Science Photo Library, **p11 b** Martyn F. Chillmaid; **p13** Jim Varney/Science Photo Library; **p14** J. C. Revy/Science Photo Library; **p15** Eric and David Hosking/Corbis UK Ltd.; **p16 t** Rick Price/Corbis UK Ltd., **p16 b** Simon Fraser/Science Photo Library; **p18 t** James Fraser/Rex Features, **p18 b** Alfred Pasieka/Science Photo Library; **p20** Martin Dohrn/Science Photo Library; **p21 t** John Cleare Mountain Camera, **p21 b** Dave Bartruff/Corbis UK Ltd.; **p22 tl** Focus Group, LLC/Alamy, **p22 bl** Stock Connection Inc./Alamy, **p22 tc** Corbis UK Ltd, **p22 bc** David Stoecklein/Corbis UK Ltd, **p22 tr** Wartenberg/Picture Press/Corbis UK Ltd, **p22 br** Owen Franken/Corbis UK Ltd; **p23** Mark Clarke/Science Photo Library; **p24 tl & tc & tr** Zooid Pictures, **p24 b** Corbis UK Ltd; **p27** Professors P. M. Motta & F. M. Magliocca/Science Photo Library; **p28** Martyn F. Chillmaid; **p30 t** Blair Seitz/Science Photo Library, **p30 b** Getty Images; **p32** Marco Cristofori/Corbis UK Ltd; **p33** Don Mason/Corbis UK Ltd; **p38 t** Andrew Lambert Photography, **p38 b** Brijesh Singh/Reuters/Corbis UK Ltd; **p40 t** Andrew Lambert Photography/Science Photo Library, **p40 b** Andrew Lambert Photography; **p41** Charles D. Winters/Science Photo Library; **p44** Herve Berthoule/Jacana/Science Photo Library; **p45** Andrew Lambert Photography; **p46 tl** William B. Jensen/Oesper Collections: University of Cincinnati, **p46 bl** Martyn F. Chillmaid, **p46 r** Dept. of Physics, Imperial College/Science Photo Library; **p47** Roger Ressmeyer/Corbis UK Ltd; **p49** David Parker/Science Photo Library; **p50** Dept. of Physics/Imperial College/Science Photo Library; **p54 tl** Arnold Fisher/Science Photo Library, **p54 tr & tcl** Andrew Lambert Photography, **p54 c** Josè Manuel Sanchis Calvete/Corbis UK Ltd, **p54 b** Dirk Wiersma/Science Photo Library, **p54 tcr** Charles D. Winters/Science Photo Library; **p55 t** Andrew Lambert Photography, **p55 b** Charles D. Winters/Science Photo Library; **p56** Science Photo Library; **p60** Charles D. Winters/Science Photo Library; **p61 t** Andrew Lambert Photography, **p61 c** Arnold Fisher/Science Photo Library, **p61 b** Martyn F. Chillmaid; **p66** Magrath Photography/Science Photo Library; **p68 l** NASA, **p68 r** Sutton Motorsport Images; **p69** Peter Turnley/Corbis UK Ltd; **p70 l** David Parker/Science Photo Library, **p70 r** Andrew Syred/Science Photo Library; **p74** (all) Zooid Pictures; **p75 tl & tr** Essex Police, **p75 bl** Mark Seymour/Oxford University Press, **p75 br** Adam Hart-Davis/Science Photo Library; **p78** Zooid Pictures; **p80** Empics; **p81** (all) Cambridge Science Media; **p82** EuroNCAP Partnership; **p85** Corbis UK Ltd; **p86** Tom Stewart/Corbis UK Ltd; **p88** Alton Towers; **p94tl & tc** Oxford Scientific Films/photolibrary.com, **p94 bl** Michael & Patricia Fogden/Corbis UK Ltd., **p94 bc & tr & br** Corel/Oxford University Press; **p96 t** Edelmann/Science Photo Library, **p96 c** Alexander Tsiaras/Science Photo Library, **p96 b** Science Photo Library; **p97 t** M.I. Walker/Science Photo Library, **p97 tr** Bob Rowan/Progressive Image/Corbis UK Ltd., **p97 br** Leo Batten/Frank Lane Picture Agency/Corbis UK Ltd; **p98 t** Joe McDonald/Corbis UK Ltd, **p98 b** Astrid & Hanns-Frieder Michler/Science Photo Library; **p99 t** M J Higginson/Science Photo Library, **p99 b** Martyn F. Chillmaid/Science Photo Library; **p101** Rob Lewine/Corbis UK Ltd.; **p103 l** Francis Leroy/Biocosmos/Science Photo Library, **p103 r** Corel/Oxford University Press; **p104 tl** Carl & Ann Purcell/Corbis UK Ltd., **p104 cl & cr** Holt Studios International, **p104 bl** CNRI/Science Photo Library, **p104 tr** Martin Harvey/Corbis UK Ltd, **p104 br** Dr Jeremy Burgess/Science Photo Library; **p106 t** Science Photo Library/Science Photo Library, **p106 b** A. Barrington Brown/Science Photo Library; **p108 l** Bob Gibbons/Holt Studios International, **p108 r** Helen Mcardle/Science Photo Library; **p109 tl** Anthony Bannister/Gallo Images/Corbis UK Ltd., **p109 bl** Russ Munn/Agstock/Science Photo Library, **p109 tr** Marko Modic/Corbis UK Ltd., **p109 br** foodfolio/Alamy; **p110 t** Biophoto Associates/Science Photo Library, **p110 c** Dr Yorgos Nikas/Science Photo Library, **p110 b** Dr Y. Nikas/Phototake Inc/Alamy; **p111** Edelmann/Science Photo Library; **p112** Mauro Fermariello/Science Photo Library; **p114** Div. of Computer Research & Technology/National Institute of Health/Science Photo Library; **p116 t** Oxford Scientific Films/photolibrary.com, **p116 b** John Kaprielian/Science Photo Library; **p125 tc & bc** Nina Towndrow/Nuffield Curriculum Centre, **p125 t & b** Adam Hart-Davis/Science Photo Library; **p126** Galen Rowell/Corbis UK Ltd; **p128 tl** George Bernard/Science Photo Library, **p128 bl** Roberto de Gugliemo/Science Photo Library, **p128 c & r** José Manuel Sanchis Calvete/Corbis UK Ltd; **p129 l** Arnold Fisher/Science Photo Library, **p129 r** Peter Falkner/Science Photo Library; **p130 tl** Sinclair Stammers/Science Photo Library, **p130 bl** Left Lane Productions/Corbis UK Ltd, **p130 r** John Mead/Science Photo Library; **p131** Mike Widdowson; **p132 t & b** Nina Towndrow/Nuffield Curriculum Centre; **p133 t** Nina Towndrow/Nuffield Curriculum Centre, **p133 b** Sciencephotos/Alamy; **p136 t** Layne Kennedy/Corbis UK Ltd,

p136 b James L. Amos/Corbis UK Ltd; **p137** Kevin Fleming/Corbis UK Ltd; **p142 tl** Alexis Rosenfeld/Science Photo Library, **p142 bl** H. David Seawell/Corbis UK Ltd, **p142 tr** John Van Hasselt/Sygma/Corbis UK Ltd., **p142 br** Charles E. Rotkin/Corbis UK Ltd.; **p145** Nik Wheeler/Corbis UK Ltd; **p150 t** Kent Wood/Science Photo Library, **p150 b** Charles D. Winters/Science Photo Library; **p153** Ron Chapple/Thinkstock/Alamy; **p157** Anthony Redpath/Corbis UK Ltd; **p158** Zooid Pictures; **p161 t** The Image Bank/Getty Images, **p161 b** sciencephotos/Alamy; **p162** Ralph Krubner/Stock Connection, Inc./Alamy; **p163** Peter Dazeley/Alamy; **p167** Vic Singh Studio/Alamy; **p168 t** Martyn F. Chillmaid/Science Photo Library, **p168 b** Andrew Lambert Photography/Science Photo Library; **p170** Barry Batchelor/PA Photos; **p172 t** Anthony Vizard/Eye Ubiquitous/Corbis UK Ltd, **p172 b** Scottish Power; **p173 l** British Energy, **p173 c** Nicholas Bailey/Rex Features, **p173 r** Scottish Power; **p178 bl** Astrid & Hanns-Frieder Michler/Science Photo Library, **p178 br** Eye of Science/Science Photo Library, **p178 t** Dennis Kunkel/PHOTOTAKE Inc./Alamy; **p179 tl** Jeff Rotman/Nature Picture Library, **p179 bl** Manfred Danegger/NHPA, **p179 tr** Tobias Bernhard/Oxford Scientific Films/photolibrary.com, **p179 br** Clive Druett/Papilio/Corbis UK Ltd; **p180 l** Sheila Terry/Science Photo Library, **p180 c** Owen Franken/Corbis UK Ltd, **p180 r** Laura Dwight/Corbis UK Ltd; **p181 l** BSIP Astier/Science Photo Library, **p181 c** Jennie Woodcock/Reflections Photolibrary/Corbis UK Ltd, **p181 r** Corbis UK Ltd.; **p182 t & b** Adam Hart-Davis/Science Photo Library; **p184 l** Jeffrey L. Rotman/Corbis UK Ltd, **p184 c** Eric and David Hosking/Corbis UK Ltd, **p184 r** Photolibrary.com; **p187 tl** Lawrence Manning/Corbis UK Ltd, **p187 tc** Corbis UK Ltd, **p187 tr** Sally and Richard Greenhill/Alamy, **p187 b** Open University; **p188 t** Corbis UK Ltd, **p188 b** Sipa Press/Rex Features; **p191 t** S. Kramer/Custom Medical Stock Photo/Science Photo Library, **p191 b** Mark Lythgoe and Steve Smith/ Wellcome Trust; **p192** Steve Bloom Images/Alamy; **p193** Anthony Bannister/ Gallo Images/Corbis UK Ltd; **p195** Jerry Wachter/Science Photo Library; **p196** Karen Kasmauski/Corbis UK Ltd; **p197** Roger Ressmeyer/ Corbis UK Ltd; **p200** Sipa Press/Rex Features; **p201** Clarissa Leahy/ Photofusion Picture Library; **p206** Maximilian Stock Ltd/Science Photo Library; **p207 l** Geoff Tompkinson/Science Photo Library, **p207 r** William Taufic/Corbis UK Ltd; **p208 tl** Dave Bartruff/Corbis UK Ltd, **p208 bl & br** Martyn F. Chillmaid, **p208 tr** Andrew Lambert Photography; **p209 l & c** Martyn F. Chillmaid, **p209 r** Lurgi Metallurgie/Outokumpu; **p210** Zooid Pictures; **p211** Andrew Lambert Photography/Science Photo Library; **p214 tl & bl & br** Zooid Pictures, **p214 tr** Richard Megna/Fundamental Photos/Science Photo Library; **p215** BSIP/Beranger/Science Photo Library; **p216** Martyn F. Chillmaid; **p217** (all) Martyn F. Chillmaid; **p218** Peter Bowater/Alamy; **p222** Gary Banks/BP Saltend; **p226** Holt Studios International; **p229** Sidney Moulds/Science Photo Library; **p234** Robert Shaw/Corbis UK Ltd; **p238 t & b** Peter Gould; **p239 t & b** Peter Gould, **p239 c** Scottish Power; **p243** Amanda Friedman/The Image Bank/Getty Images; **p244** D. Boone/Corbis UK Ltd; **p246 l** Pascal Goetgheluck/Science Photo Library, **p246 r** JPL/NASA; **p247** Lester V. Bergman/Corbis UK Ltd; **p248 t** Geoff Tompkinson/Science Photo Library, **p248 b** Shout; **p249** Robin Millar; **p250** Victor De Schwanberg/Science Photo Library; **p251** Zooid Pictures; **p253** BBC Points West/BBC Bristol; **p255** Dr Jeremy Burgess/Science Photo Library; **p256 t** Alfred Pasieka/Science Photo Library, **p256 b** Pure Digital; **p257** A. Tovy/Robert Harding Picture Library Ltd/Alamy.

Molymod Molecular Models supplied by Spiring Enterprises Ltd.

Illustrations by IFA Design, Plymouth, UK and Clive Goodyer.

OXFORD
UNIVERSITY PRESS

Great Clarendon Street, Oxford OX2 6DP

Oxford University Press is a department of the University of Oxford.
It furthers the University's objective of excellence in research, scholarship,
and education by publishing worldwide in

Oxford   New York

Auckland   Cape Town   Dar es Salaam   Hong Kong   Karachi
Kuala Lumpur   Madrid   Melbourne   Mexico City   Nairobi
New Delhi   Shanghai   Taipei   Toronto

With offices in

Argentina   Austria   Brazil   Chile   Czech Republic   France   Greece
Guatemala   Hungary   Italy   Japan   Poland   Portugal   Singapore
South Korea   Switzerland   Thailand   Turkey   Ukraine   Vietnam

British Library Cataloguing in Publication Data

Data available

ISBN-13: 978-0-19-915044-1

10 9 8 7 6 5

Typeset by IFA Design, Plymouth, UK

Printed at Cayfosa Quebecor, Spain

Project Team acknowledgements

These resources have been developed to support teachers and students undertaking a new OCR suite of
GCSE science specifications, *Twenty First Century Science*.

Many people from schools, colleges, universities, industry, and the professions have contributed to the production
of these resources. The feedback from over 75 Pilot Centres was invaluable. It led to significant changes to the
course specifications, and to the supporting resources for teaching and learning.

The University of York Science Education Group (UYSEG) and Nuffield Curriculum Centre worked in partnership
with an OCR team led by Mary Whitehouse, Elizabeth Herbert, and Emily Clare to create the specifications,
which have their origins in the *Beyond 2000* report (Millar & Osborne, 1998) and subsequent Key Stage 4
development work undertaken by UYSEG and the Nuffield Curriculum Centre for QCA. Bryan Milner and
Michael Reiss also contributed to this work, which is reported in: *21st Century Science GCSE Pilot Development:
Final Report* (UYSEG, March 2002).

Sponsors

The development of *Twenty First Century Science* was made possible
by generous support from:

- The Nuffield Foundation
- The Salters' Institute
- The Wellcome Trust

THE SALTERS' INSTITUTE

The Nuffield Foundation

wellcome trust